Advances
in
Genetics

Advances in Genetics, Volume 58

Serial Editors

Jeffery C. Hall
Waltham, Massachusetts

Jay C. Dunlap
Hanover, New Hampshire

Theodore Friedmann
La Jolla, California

Veronica van Heyningen
Edinburgh, United Kingdom

Volume 58

Advances in Genetics

Edited by

Jeffery C. Hall
Department of Biology
Brandeis University
Waltham, Massachusetts

AMSTERDAM • BOSTON • HEIDELBERG • LONDON
NEW YORK • OXFORD • PARIS • SAN DIEGO
SAN FRANCISCO • SINGAPORE • SYDNEY • TOKYO
Academic Press is an imprint of Elsevier

ELSEVIER

Academic Press is an imprint of Elsevier
525 B Street, Suite 1900, San Diego, California 92101-4495, USA
84 Theobald's Road, London WC1X 8RR, UK

This book is printed on acid-free paper. ∞

For information on all Elsevier Academic Press publications
visit our Web site at www.books.elsevier.com

ISBN-13: 978-0-12-373882-0
ISBN-10: 0-12-373882-2

PRINTED IN THE UNITED STATES OF AMERICA
07 08 09 10 9 8 7 6 5 4 3 2 1

Contents

v

4 Mutational Analysis of the Ribosome   89
Kathleen L. Triman

5 Application of Genomics to Molecular Breeding of Wheat and Barley   121
Rajeev K. Varshney, Peter Langridge, and Andreas Graner

Contributors

Numbers in parentheses indicate the pages on which the authors' contributions begin.

Alexander J. R. Bishop (67) Department of Cellular and Structural Biology, Children's Cancer Research Institute, University of Texas Health Science Center San Antonio, San Antonio, Texas 78229

Andreas Graner (121) Institute of Plant Genetics and Crop Plant Research (IPK), D06466 Gatrersleben, Germany

Christian Heintzen (25) Faculty of Life Sciences, University of Manchester, Manchester M13 9PT, United Kingdom

Pierre Hutter (1) Division of Genetics, Institut Central des Hôpitaux Valaisans, Avenue Grand-Champsec 86, 1951 Sion, Switzerland

Peter Langridge (121) Australian Centre for Plant Functional Genomics (ACPFG), University of Adelaide, Waite Campus, PMB 1 Glen Osmond, SA 5064, Australia

Yi Liu (25) Department of Physiology, University of Texas Southwestern Medical Center, Dallas, Texas 75390

Ramune Reliene (67) Department of Pathology, Geffen School of Medicine, Los Angeles, California 90024

Robert H. Schiestl (67) Department of Pathology and Department of Radiation Oncology, Geffen School of Medicine, Los Angeles, California 90024; Department of Environmental Health, School of Public Health, Los Angeles, California 90024

Kathleen L. Triman (89) Department of Biology, Franklin and Marshall College, Lancaster, Pennsylvania 17604

Rajeev K. Varshney (121) International Crops Research Institute for the Semi-Arid Tropics (ICRISAT), Patancheru 502 324, A.P., India

Rapidly Evolving *Rab* GTPase Paralogs and Reproductive Isolation in *Drosophila*

Pierre Hutter

Division of Genetics, Institut Central des Hôpitaux Valaisans
Avenue Grand-Champsec 86, 1951 Sion, Switzerland

ABSTRACT

Alterations at the X-linked *Hmr* gene of *Drosophila melanogaster* can fully restore viability and partially restore fertility in hybrid flies from crosses between *D. melanogaster* and any of its three most closely related species. Although more than one gene is expected to be involved in these barriers to reproduction, a single DNA-binding protein was recently identified as HMR. The *Hmr* gene was shown to evolve unusually fast, a feature that supports its role in causing genetic incompatibility in a hybrid genotype. The current treatment of hybrid genetics focuses not only on *Hmr* but also on the *Rab9D* gene, which lies only 1 kb from *Hmr*. *Rab9D* is proposed also to influence hybrid viability. This gene has remained tightly linked to *Hmr* for about 10 million years, but it has diverged even more than *Hmr* with regard to *D. melanogaster* and its most closely related species.

Advances in Genetics, Vol. 58
0065-2660/07 $35.00
DOI: 10.1016/S0065-2660(06)58001-0

Furthermore, the 197-amino acid RAB9D protein contains four amino acid substitutions in the *D. melanogaster*-rescuing mutant *Hmr1*. *Rab9D* is shown to have evolved under very strong positive selection and to be the most recent member of a cluster of six paralogs that encode small RAB GTPases. Four of the six paralogs are unique to *D. melanogaster* in which they have diverged considerably, their encoded proteins sharing less than 50% amino acid identities with proteins from their orthologs in the closest species. Only two *Rab* orthologs are present in these sibling species and none is present in the genomes of more distantly related *Drosophila* species. Rapidly evolving *Rab* paralogs near the *Hmr* locus probably developed functional specialization of redundant proteins involved in trafficking macromolecules between cytoplasm and nucleus. Positive selection acting on duplicates of these *Rab* genes appears to participate in reproductive isolation.   © 2007, Elsevier Inc.

I. INTRODUCTION

It has often been argued that Darwin did not solve the issue of "The Origin of Species" because small genetic variations can only account for differences between populations within a species and not account for divergence between species by macroevolution. Findings discussed here support the view that gene duplication, an event relevant to macroevolution, together with classical Darwinian microevolution, may be instrumental in the process of reproductive isolation. Indeed, a high rate of evolution of genes involved in reproductive isolation can be facilitated when these involve paralogs that are subject to less selective constraints than the parental form.

  According to the biological species concept formulated by Mayr (1943), the creation of a new species requires that members of a population build up their own gene pool after becoming reproductively isolated from other members of the parental species. In organisms produced by crosses between populations in the state of incipient speciation, genetics posits that barriers to gene flow, such as lethality or sterility, result from genetic factors that exhibit functional incompatibilities. In hybrids between sibling species that have since long become distinct biological entities, it is difficult to identify, *a posteriori*, which of the genes associated with hybrid incompatibility (HI) actually reflect the primary factors that were instrumental to the initiation of speciation.

  While both Cordon and Kosswig proposed as early as 1927 the concept of genetic HI in the fish genus *Xiphophorus*, entry points for experimental studies on this major biological issue only emerged over the last 25 years (Cordon, 1927; Kosswig, 1927). A handful of observations have indicated that single genetic changes can drastically influence reproductive isolation between most closely related species. The first mutations that can break through reproductive isolation were reported in organisms amenable to genetic analysis, such as *Lhr* and *mhr* in *Drosophila simulans* (Sawamura *et al.*, 1993a; Watanabe, 1979), as well as *Hmr* and *Zhr* in *D. melanogaster* (Hutter and Ashburner, 1987; Hutter *et al.*, 1990; Sawamura *et al.*, 1993b). These genetic changes are sufficient to restore full viability in otherwise completely inviable hybrids, and some of these alterations also partly restore female fertility. Similarly, mutations in otherwise fully fertile flies were found to play a significant role in causing sterility of hybrids between *Drosophila* species (see Wu and Ting, 2004 for a review). Because genetic alterations responsible for these rescues have no phenotypic effect within a given species, they are thought to reflect incompatibilities in hybrid genotypes having to do with genes that have diverged functionally.

As Coyne and Orr (2004) emphasized, the first molecular studies of these genes strongly suggested that they represent the actual genes that cause the death or the sterility of hybrids rather than being second-site suppressors that might ameliorate effects of the loci causing hybrid problems. Thus, these genetic alterations influencing HI are our best candidates to explore the scenario of reproductive isolation based on a minimum of two genes, as originally envisioned by Dobzhansky (1937) and Müller (1940). Noteworthy, in *D. melanogaster* all mutations that rescue otherwise lethal hybrids were found in genes of the X chromosome, and previous studies in other *Drosophila* species have also indicated a major effect of the X chromosome on hybrid inviability (reviewed in Hutter, 1997). Genes that cause HI as a result of positive selection are predicted to be often X-linked, as recessive X-linked alleles can become fixed more quickly than autosomal genes and are likely to evolve rapidly (Charlesworth *et al.*, 1987).

Under laboratory conditions, *D. melanogaster* can hybridize with its three most closely related species, *D. simulans*, *D. mauritiana*, and *D. sechellia* (hereafter referred to as the sibling species), even though the latter species have diverged from an ancestor of *D. melanogaster* approximately 2–3 million years (Myr) ago (Lachaise *et al.*, 1988). However, crosses between *D. melanogaster* females and males of any of its three sibling species produce hybrid daughters that are viable but sterile at low temperatures and lethal at high temperatures. Hybrid sons invariably die as late larvae or pseudopupae and never metamorphose (Hadorn, 1961; Hutter *et al.*, 1990; Lachaise *et al.*, 1988; Sturtevant, 1920). Three *D. melanogaster* mutants—*Hmr1*, *In(1)AB*, and *Df(1)EP307-1-2*—were found to rescue otherwise lethal hybrids, and all three mutations map to cytological interval 9D–9E in the middle of the X chromosome (Barbash *et al.*, 2000, 2004b; Hutter and Ashburner, 1987; Hutter *et al.*, 1990). The first two mutations not only rescue lethal hybrids, but also contribute to restore fertility in otherwise sterile hybrid females (Barbash and Ashburner, 2003).

During recent years, three studies have addressed the molecular basis of the genetic factors capable to suppress the invariant lethality of hybrids. The first report postulated that a cluster of six paralogs in *D. melanogaster*, which encode RAB GTPase proteins, is involved in HI between members of the *D. melanogaster* subgroup species (Hutter, 2002). Sexually antagonistic coevolution between the *Rab* paralogs and extranuclear components were hypothesized to result in fast evolution of genes involved in vesicle trafficking and cell signaling. The second molecular study, based on transgenesis experiments, identified a single gene as *Hmr* (Barbash *et al.*, 2003), lying in the middle of the above cluster of six *Rab* paralogs. *Hmr* encodes a regulatory protein with homology to a family of MADF- and MYB-related DNA-binding transcriptional regulators and was shown to be evolving particularly fast as a result of strong positive selection (Barbash *et al.*, 2003, 2004a). Indeed, an unusually high average divergence rate of 7.7% for nonsynonymous nucleotide substitutions [leading to amino acid (AA) replacements] was observed in HMR protein between *D. melanogaster* and its sibling species. Nonetheless, in *In(1)AB* and *Df(1)EP307–1-2 D. melanogaster* mutants which rescue the same interspecific hybrids as *Hmr1* does, no causative mutation has yet been attributed to *Hmr*. This gene appears normally transcribed in both above-mentioned mutants by Northern analysis (Barbash *et al.*, 2003). The third molecular study on HI in *Drosophila*, based on complementation mapping of HI genes between *D. melanogaster* and *D. simulans*, identified a gene called *Nup96*, on the third chromosome of *D. simulans* (Presgraves *et al.*, 2003), whose product acts at the nuclear pore complex.

     In the current treatment of this subject, I further investigate the role of duplication and variation of *Rab* GTPase paralogs mapping near *Hmr*, with respect to viability of hybrids involving members of the *D. melanogaster* species subgroup. Four of the six *Rab* paralogs near *Hmr* are shown to have diverged more than *Hmr*, between *D. melanogaster* and its sibling species, and to have evolved under strong positive selection. Moreover, a mutation which inactivates the *Rab* paralog which lies 1 kb from *Hmr*, is shown to reduce HI, whereas the mutation has a deleterious effect in *D. melanogaster*. Putative functions of the proteins encoded by the *Rab* genes in the vicinity of *Hmr* are related to proteins acting at the nuclear pore complex which, as just mentioned, is involved in HI (Presgraves *et al.*, 2003). Since the nuclear pore complex is the sole site of cytonuclear trafficking of RNAs and proteins, I argue that the lethality of hybrids between members of the *D. melanogaster* species subgroup may be related to impaired nucleocytoplasmic transport. In this view, the confluence of signaling mechanisms, epithelial morphogenesis, and cytoskeleton dynamics would be imbalanced in hybrids. The two microtubule-based morphogenetic processes of protein sorting and generation of transport carriers are regulated by members of the same signaling pathways, since microtubules serve as tracks that not only facilitate targeted vesicular transport, but also play a major role in promoting protein sorting (Musch, 2004). Besides, the participation of RAB GTPase proteins in disposing of malfolded newly synthesized proteins as well as in the degradation of signaling receptors may become critical in a hybrid genotype.

II. EXPERIMENTAL APPROACHES

D. melanogaster flies carrying the *PBac{WH}Rab9D*[f06390] transposon were obtained from the Bloomington, Indiana, stock center. *D. mauritiana* C164.1 flies, collected from the Rivière Noire, Mauritius, were obtained from the Department of Genetics (Cambridge, UK). Wild-type *D. melanogaster* of the Canton-S type were obtained from the Department of Zoology and Animal Biology (Geneva, Switzerland), and Saxon-04 wild-type flies were collected by me in Saxon, Switzerland. Crosses between three *D. melanogaster* females and five *D. mauritiana* males were set up at 24°C for 48 h. After removing the parents, cultures were shifted to 18°C. The numbers of larvae, male pseudopupae, and female adults were only scored from vials which contained between 50 and 150 larvae. The phrase "dead pharates" refers to flies which died within their pupal case but whose eye pigmentation could be seen. Viability of *D. melanogaster* flies carrying the *PBac{WH}Rab9D*[f06390] transposon was assessed in flies maintained at 20°C.

Molecular analyses were carried out after obtaining genomic DNA from a given group of 30 flies (using the purification system from Puregene, Gentra, Minnesota). Sequence analysis was carried out in each case from three PCR products obtained from genomic DNA, and in one case from plasmid p94 (pBSIIKS+, kindly provided to me by D. Barbash), to analyze synteny. PCR conditions, sequences of primers used for DNA amplification and sequencing are available on request to PH (e-mail: pierre.hutter@ichv.ch). Sequences were aligned using MACAW (Multiple Alignment Construction and Analysis Workbench) software (Schuler *et al.*, 1991). D_n and D_s values were calculated for each pairwise comparison of *Rab*-coding sequences, by the maximum-likelihood method implemented in PAML3.15 (Yang, 2004). The same program package was used to perform a phylogenetic analysis in an attempt to build an evolutionary tree of the *Rab* genes under study. (I am very grateful to H. Kaessmann for help with PAML analysis.)

III. INFLUENCES OF *Rab9D* ON VIABILITY OF HYBRIDS BETWEEN *D. MELANOGASTER* AND ITS SIBLING SPECIES

Six paralogous *Rab*-encoded GTPases genes from cytological region 9 on the X chromosome (hereafter referred to as *6paRab9*) were identified in the cytological region 9C-9F of *D. melanogaster* (Hutter, 2002; Hutter and Karch, 1994). These genes were later reported as *RabX2*, *Rab9D*, CG 9807 (here named *Rab9Db*), CG 32673 (here named *Rab9E*), CG 32671 (here named *Rab9Fa*), and CG 32670 (here named *Rab9Fb*) in the 2003 release of the *D. melanogaster* database (http://flybase.bio.indiana.edu.). The genomic positions of the *6paRab9* around the *Hmr* locus are shown in Fig. 1.1. All six *Rab* genes have retained an

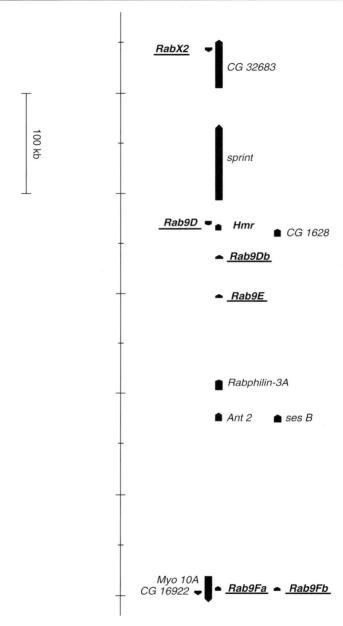

Figure 1.1. Cluster of six *Rab* GTPases genes (*6paRAB9*) in cytological region 9C-9F of the X chromosome of *D. melanogaster*. Not all genes present in the interval are shown. The *6paRAB9* genes comprised within about 550 kb of genomic DNA are underlined, and nine other genes of particular interest are reported (see text).

open reading frame as a single exon and share strong similarity not only through their coding region but also through their 5′ and 3′ untranslated regions, as well as through up to 600 base pairs (bp) of their flanking genomic sequences. Yet important shuffling of these regions has occurred between paralogs, as illustrated by alignments of sequences from *Rab9D* and *Rab9Db*, which are very similar at their 3′ ends but differ strongly at their 5′ ends. It must be emphasized that large parts of a 360-bp sequence are found about 280-bp 5′ of the coding sequences of all *6paRAB9* genes but *Rab9D*, as well as about 460-bp 3′ of the coding sequences of four of these genes. Interestingly, large parts of this sequence are also present at more than 300 sites throughout the X chromosome, whereas they are found at less than 20 sites on the autosomes. One X-linked site lies 30-bp 3′ of CG 9806, a gene adjacent to *Rab9Db* and involved in proteolysis. Moreover, the 15 repeats that share more than 90% identities with the copy present 3′ of *Rab9D* are all clustered around the *6paRAB9*, between chromosomal regions 9C and 9F. More than 200 similar sequences are present on the X chromosomes of each of the 5 species of the *melanogaster* subgroup species, but no homologous sequence is present in published genomes from other species of *Drosophila*. The possibility that the repeat influences chromatin structure and behaves as chromatin domains insulator is worth considering. The sequence contains a few weak binding motifs of the transcription factors BEAF-32, HSP70, and KNIRPS, which are associated with histone proteins required for heterochromatin state, and which are also involved in the degradation of cytoplasmic unfolded proteins. This might further point to a link with the RAB GTPase relationship with the proteasome.

As will be discussed later, the two paralogs *RabX2* and *Rab9Fb* may have retained an ancestral function. These two genes as well as *Rab9Fa* are located at each edge of the *6paRAB9* cluster (Fig. 1.1), where they are nested within two large genes CG 32683 and Myo10A. When considering the genomic distance separating all but two *Rab* paralogs, and that these genes are intron-less, the derived duplicates are likely to have been generated by retrotransposition rather than by unequal crossing over. The latter process is predicted to generate duplications of large gene regions including other genes, which is clearly not the case for the *6paRAB9*. In addition, a BLAST search indicated that 64 bp (16%) the first 400-bp downstream of *Rab9D* harbor four sequences that are 100% homologous with transposable elements.

A. Inactivation of *Rab9D* reduces hybrid incompatibility

Three lines of evidence suggest an implication of 6paRab9 in HI between members of the *D. melanogaster* subgroup species. First, a *piggyBac* transposon-induced mutation, which knocks out *Rab9D*, was found to reduce HI. The site of insertion in *Rab9D* of the lepidopteran *transposon PBac{WH}Rab9D*[f06390]

(Thibault *et al.*, 2004) was first clarified by genomic sequencing, which placed it at the TTAA target nucleotides 23–26 from the A within the gene's translation-start codon. Inactivating *Rab9D* in *D. melanogaster* improves, at the larval and the pupal stages, the fate of hybrids of both sexes flies obtained from the cross between *D. melanogaster* females and *D. mauritiana* males. As discussed below, the mutation has the opposite effect in *D. melanogaster*, resulting in a significant reduction in viability of both sexes. It is well established that the cross between wild-type *D. melanogaster* females and *D. mauritiana* males produces hybrid sons which consistently die as late third instar larvae, or as pseudopupae that soon exhibit patches of necrotic tissues (Hutter *et al.*, 1990; Lachaise *et al.*, 1988). These hybrid males are typically seen as very sluggish larvae at the surface of the food, several days after their sisters have emerged from their pupal case. Hybrid survival is better at low temperatures, but even at 18 °C not all hybrid daughters from the above cross survive. The proportion of hybrid females that fail to reach the adult stage was shown previously to increase with temperature (Barbash *et al.*, 2000). By considering the strong homology between *Rab9D* and its three closest paralogs reported below, inactivating only one of the four paralogs is not predicted to drastically alter hybrid viability on its own. Nonetheless, a clear-cut difference in the fate of both male and female hybrids, respectively hemizygous and heterozygous for the mutation, was observed (Table 1.1A and 1.1B).

With regard to crosses between *D. melanogaster* females and *D. mauritiana* males, 16 were performed with $Rab9D^+/Rab9D^+$ wild-type females (8 using

Table 1.1A. Hybrid Daughters Recovered from Interspecific Crosses Between *D. melanogaster* Females and *D. mauritiana* Males

Cross	♀ Alive	♀ Dead eclosed	♀ Dead pharate	Total	Viability (%)
D. melanogaster Canton-S ♀ × *D. mauritiana*-C.164.1 ♂	1012	24	71	1107	91.4
D. melanogaster Saxon-04 ♀ × *D. mauritiana*-C.164.1 ♂	851	16	49	916	92.9
D. melanogaster Oregon-R ♀ × *D. mauritiana*-C.164.1 ♂	76[a]	3[a]	2[a]	81[a]	93.8[a]
D. melanogaster PBac{WH}Rab9D^f06390/PBac{WH}Rab9D^f06390 ♀ X *D. mauritiana*-C.164.1 ♂	1686	0	1	1687	100

[a]Data from Barbash *et al.* (2000).

Table 1.1B. Hybrid Sons Recovered from Interspecific Crosses Between *D. melanogaster* Females and *D. mauritiana* Males

Cross	♂ Pseudopupa	%	♂ Third instar larva	♂Adult	♀Adult
D. melanogaster Canton-S ♀ × *D. mauritiana*-C.164.1♂	149	13.5	716[a] counted	0	1107
D. melanogaster Saxon-04 ♀ × *D. mauritiana*-C.164.1 ♂	113	12.3	593[a] counted	0	916
PBac{WH}Rab9D[f06390]/PBac {WH}Rab9D[f06390] ♀ × *D. mauritiana*-C.164.1 ♂	429	25	810[b] counted	1[c]	1687

[a]Brown to black color after 1–2 weeks.
[b]Light brown color after 3–4 weeks.
[c]XO male resulting from nondisjunction.

Canton-S mothers and 8 using Saxon-04 mothers) and 14 with *Rab9D⁻/Rab9D⁻* females. As shown in Table 1.1A, between 6.2% and 8.6% of daughters inheriting a wild-type copy of *Rab9D* from their mothers died either as pharates or as eclosed adults, consistent with earlier findings (Barbash *et al.*, 2000). In contrast, only one of the 1687 daughters recovered from the cross with mothers carrying *Rab9D⁻* died. Similarly, 429 (ca. 25%) of their brothers carrying *Rab9D⁻* were able to form a pseudopupa, which did not exhibit necrotic tissue patches for up to 5 weeks (Table 1.1B), that is, about 80 days after egg laying (AEL). The remaining hybrid sons were found as stiff brown third instar larvae with no sign of necrotic tissue, and some larvae were still moving in the food 85 days AEL. This is different from the fate of hybrid males from the same interspecific cross, but carrying a wild-type copy of *Rab9D*. Only 262 out of 2023 (13%) *Rab9D⁺* male larvae formed pseudopupae, while their brothers died as brown dark larvae (Table 1.1B). Moreover, *Rab9D⁺* hybrid males typically exhibited necrotic tissues within 2 weeks after pupariation (about 50 days AEL). All remaining larvae that could be counted inside the food were found dead as stiff black third instar larvae 70 days AEL.

To determine the effect of the transposon in nonhybrid flies, survival of *Rab9D⁻ D. melanogaster* adults was also investigated, which indicated that mutant larvae are significantly less viable than wild-type larvae. In progeny from a cross between females homozygous for the transposon and Canton-S males, 234 males and 273 females (still heterozygous for the transposon) were counted as adults (Table 1.2). The same cross, but using males from the wild-type strain Saxon-04, yielded 364 males and 456 females. This corresponds to a significant departure from a 1:1 sex ratio (Table 1.2), indicating that the transposon has

Table 1.2. Effect of Transposon-Induced Inactivation of the *Rab9D* Gene in *D. melanogaster*

Cross	♂ +/Y	♂ Rab9D⁻/Y	♀ +/+	♀ Rab9D⁻/Rab9D⁻	♀ Rab9D⁻/+	Departure[a] from 1:1 SR
Rab9D⁻ᵇ/Rab9D⁻ ♀ × Canton-S ♂	–	234	–	–	273	$p < 0.02$
Rab9D⁻/Rab9D⁻ ♀ × Saxon-04 ♂		364	–	–	456	$p < 0.001$
Rab9D⁻/+♀ × *Rab9D⁻/Y* ♂	190	146	–	150	185	$p < 0.005$
Rab9D⁻/Rab9D⁻ ♀ × *Rab9D⁻/Y* ♂	–	143	–	149	–	NS
Saxon-04 ♀ × Saxon-04 ♂	207	–	211	–	–	NS

[a]Using a chi-square test.
[b]*PBac{WH}Rab9Dᶠ⁰⁶³⁹⁰*.
SR, sex ratio; NS, not significant.

a detrimental effect in *D. melanogaster*, as opposed to its effect on improved viability of hybrids. In progeny of the above cross with Saxon-04 males, *Rab9D/⁺*daughters were crossed to *Rab9D⁻/Y* sons. As shown in Table 1.2, a departure from a 1:1 sex ratio was again observed for both sons and daughters carrying one copy of the *Rab9D⁻* allele. Table 1.2 also shows that the sex ratios of flies carrying the transposon *PBac{WH}Rab9Dᶠ⁰⁶³⁹⁰*, and of Saxon-04 flies, were regular.

In order to further investigate the causative effect of the transposon on hybrid viability, I took advantage of the fact that this insertion is tagged with the *mini-white* gene. Sons and daughters from the cross between females homozygous for the transposon and Canton-S males were crossed *inter se*, and at the next generation 10 females homozygous for the transposon were crossed individually to *D. mauritiana* males. All nine cases with hybrid progeny from these crosses fully retained the improved survival as described above, and no hybrid daughter was found dead out of 871 females. No attempt was made to excise the transposon, though only a direct comparison of excision and nonexcision strains would formally rule out any effect of the genetic background of the X chromosome carrying the transposon. Still, the probability that the observed reduction in HI was due to a concomitant genetic alteration is rather low, as only 0.3–0.5% of *piggyBac* carrying stocks are expected to carry unlinked background mutations (Thibault *et al.*, 2004). The above results support that hybrid larvae mutant for *D. melanogaster Rab9D* exhibit reduced HI, when compared to the fate of hybrids carrying a wild-type copy of this gene.

B. Fast rate of evolution of *Rab9D*

A second line of support for a role of the *6paRAB9* genes in HI comes from the observation that *Hmr1* flies carry five mutations in their *Rab9D* paralog. In *Hmr1* flies, whose *Hmr* gene is causally mutated (Barbash *et al.*, 2003), the protein encoded by *Rab9D* harbors four nonsynonymous AA replacements out of its 197 AAs (Fig. 1.2A). Noteworthy, the AAs replaced are identical to those present at the same positions in other *6paRAB9* paralogs, particularly in *Rab9E*. Further-more, BLASTING 707 nucleotides of the reference sequence, downstream of the termination codon of *Rab9D$^+$*, against a sequence from an *Hmr1* clone deposited in the Flybase (Barbash *et al.*, 2003) further revealed six nucleotide substitutions and one insertion. Fig. 1.2B summarizes the level of AA conserva-tion between the 6paRAB9 proteins from *D. melanogaster*, showing that four of the paralogs, *Rab9Fa*, *Rab9E*, *Rab9D*, and *Rab9Db*, are 98–99% identical. The last three genes are separated by only 32.8 and 38 kb of genomic DNA.

A third support for a role of the *6paRAB9* genes in HI comes from the elevated rate of divergence between the *Rab* paralogs in *D. melanogaster* and their orthologs in the sister species. In order to compare orthologs of the *Rab9D* gene between species, synteny was determined with reference to the *Hmr* gene. The *Rab* genes compared between sibling species are located between 1.5 and 2.5 kb from their respective *Hmr* ortholog. To identify the *D. melanogaster Rab9D* ortholog in *D. mauritiana* (reported under GenBank accession number AY830113), synteny with *Hmr* was established by sequencing two adjacent genes from plasmid p94 from *D. mauritiana*, which contains about 7 kb of genomic DNA encompassing *Hmr*. Interspecific comparison revealed an increasing number of copies of *Rab* genes in region 9, from the ancestral species to *D. melanogaster*. As judged from the current genome assemblies, no ortholog of the *D. melanogaster 6paRab9* genes can be identified in the genomes of *D. ananassae*, *D. pseudoobscura*, and *D. persimilis*; a single ortholog can be found in both *D. yakuba* (gb/CM000362) and *D. erecta* (scaffold_4690 freeze 1 assembly of 2.12.2005); whereas two active orthologs are present in the genomes of *D. simulans* and *D. sechellia*. In the latter species an *Rab* pseudogene contains several deletions/insertions of 1 bp, and the two active orthologs are most similar to *Rab9Fb* and *RabX2* of *D. melanogaster*. Besides, in these species no *Rab* ortholog is nested in either CG 32683 or Myo10A orthologs, as observed in *D. melanogaster* (Fig. 1.1). Moreover, *Rab9D* orthologs in the sibling species encode proteins that share between 92 and 95% identical AAs, while sharing less than 50% AAs with their most similar ortholog in *D. melanogaster*. It is notable that the maximum-likelihood estimate of the average rate of diver-gence per nonsynonymous (AA replacement) sites (D_n) between *D. melanogaster* and its sibling species is higher for *Rab9D* than for *Hmr*. As tested by the methods of Nei and Gojobori (1986), the estimate of D_n between *D. melanogaster* and *D. erecta*/ *D. yakuba* was 0.331 and 0.361, respectively, which is well above the mean value

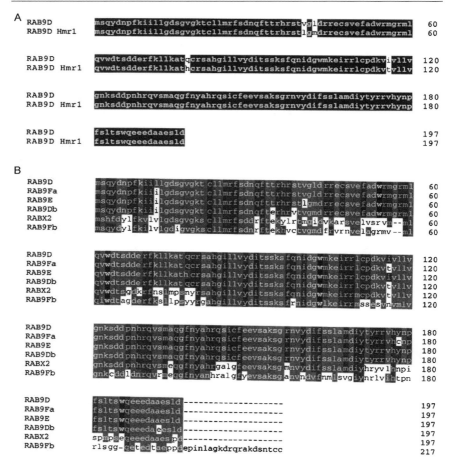

Figure 1.2. (A) Alignment of the predicted RAB9D amino acid sequences from *D. melanogaster*. The upper sequence corresponds to the *D. melanogaster* wild-type sequence and the lower sequence is from the *Hmr1* mutant, which rescues otherwise lethal hybrids, and which was found to carry a causative mutation in the adjacent *Hmr* gene (Barbash et al., 2003). The four residues that differ between the two sequences are indicated by a white background. (B) Alignment of the predicted amino acid sequences of the proteins encoded by the six *D. melanogaster Rab* paralogs (6paRAB9) in cytological region 9C-9F of the X chromosome of *D. melanogaster*. Residues most conserved across sequences are indicated by a black background.

of 0.057 found in a survey of 53 *D. erecta* genes (Bergman et al., 2002), and about twice the D_n value of 0.166 found for *Hmr* between *D. melanogaster* and *D. erecta* (Barbash et al., 2004a).

The analysis of an outgroup of _Rab9D_ orthologous sequences from the _D. melanogaster_ subgroup species allows one to address whether _Rab9D_ began to diverge rapidly after the _D. melanogaster_-sibling species divergence. For this purpose the _Rab9D_ orthologs from _D. erecta_ and _D. yakuba_, estimated to have diverged from _D. melanogaster_ between 6 and 15 Myr ago (Powell, 1997), were used. When site-specific models were applied, maximum-likelihood estimate showed strong positive selection at the _Rab9D_ gene, and these results permitted the construction of a phylogeny-based evolutionary tree as depicted in Fig. 1.3. The values shown indicate that the K_a/K_s ratios differ significantly among branches. By analyzing averaged over all sites of the sequences and potentially positively selected sites (M1 and M2 models), the results indicated strong positive selection ($2\Delta I = 51.5$, $p < 0.0001$, chi-square distribution) at _Rab9D_. When a model where all D_n/D_s ratios are free to vary along all branches was used, by far the highest rate of evolution was found on the branch leading to _D. simulans_ and _D. mauritiana_ ($D_n/D_s = 2.63$). Thus, divergence at _Rab9D_ between _D. melanogaster_ and the sibling species appears to be due to strong

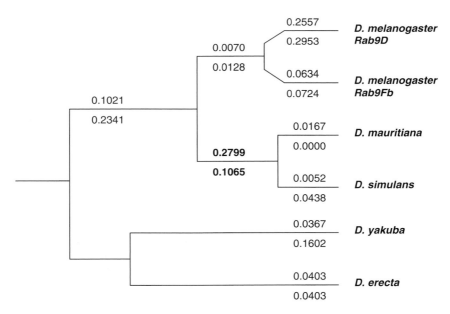

Figure 1.3. Maximum-likelihood estimates of RAB9D divergence among _Drosophila_ lineages. Estimates of the number of changes per nonsynonymous site (D_n) are shown above each lineage; and the number of changes per synonymous site (D_s) are shown below each lineage, calculated separately for each branch. The values of the very high D_n/D_s ratio (2.63) in the branch leading to _D. mauritiana_ and _D. simulans_ are indicated in bold.

positive selection in the ancestor of the sibling species, supporting a role of this gene in HI. Further analysis of potentially positively selected sites identified 11 AAs that had been positively selected with p values between 95 and 99%. Thus, in the branch leading to the sibling species $Rab9D$ ortholog appears to have been subject to strong positive selection in a relatively recent past.

This pattern of divergence is consistent with the degree of reproductive isolation observed between members of the $D.$ $melanogaster$ species subgroup. Viable hybrids of both sexes and fertile female hybrids can be produced from crosses between $D.$ $mauritiana$ and $D.$ $simulans$, but not from crosses between these two species and $D.$ $melanogaster$ (David et al., 1974, 1976).

$Rab9D$ lies about 1 kb from Hmr (Fig. 1.1) in $D.$ $melanogaster$, and the orthologs of both genes are also very closely linked in $D.$ $erecta$, $D.$ $yakuba$, and the three sibling species. This strongly suggests that the ancestors of $Rab9D$ must have been inherited together with the ancestors of Hmr most of the time since the split of the $D.$ $melanogaster$ subgroup. This event is estimated to have taken place about 6–15 Myr ago (Powell, 1997). The two genes could also have remained associated through the process of hitchhiking, which would not rely on gene function but only on closely linked locations. However, if some hitchhiking effect contributed this association, this should also be reflected as polymorphisms in neighboring genes that are not related in function.

IV. PREDICTIONS OF FUNCTIONS FOR 6paRAB9 PROTEINS

RAB proteins play major roles in signaling and in recognition of organelle identification machinery (Pfeffer, 2001). In addition to their role in controlling transport processes, such as vesicular traffic between the endoplasmic reticulum (ER) and the Golgi apparatus (Mattaj and Englmeier, 1998), RAB GTPases are implicated in the formation and organization of the cytoskeleton (Riggs et al., 2003). All six 6paRAB9 proteins are predicted to have an ATPase activity (Flybase Genome Annotators 2002–2003), as inferred from sequence similarity with RAB proteins from other organisms.

In a collaborative experiment on localization of the RAB9D protein, immunoreactivity was mainly seen in the cytoplasm (Fig. 1.4), but also within the nucleus (A. Munehiro, personal communication). It is worth pointing out that the small size of the 6paRAB proteins would allow their passive diffusion through the nuclear pore complex. RAB proteins are known to shuttle rapidly between a soluble guanosine diphosphate (GDP)-bound form and a membrane-associated GTP-bound form as they regulate vesicular trafficking (Seabra et al., 2002). By considering the wide range of function diversification of the large family of RAB GTPases, caution must be taken when interpreting sequence differences that reveal protein analogies, as even slight divergence may be sufficient to provide

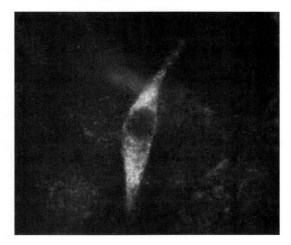

Figure 1.4. Transfection experiment showing the localization of the RAB9D protein in NIH/3T3 cells. Plasmid pEGFP-C1/N1 was fused to HA-*Rab9D* cDNA and the resulting construct was expressed in NIH/3T3 cells. HA-RAB9D was mainly localized in the cytoplasm, although it was also observed within the nucleus to some extent (reproduced with permission from A. Munehiro).

functional distinctions. Nonetheless, sequence comparison strongly suggests that the proteins encoded by four very similar *Rab* paralogs (Fig. 1.2B), including *Rab9D*, play a role in the processes of intracellular protein transport and endocytosis, as does their most similar protein, RAB13. The latter is expressed at tight junctions (Morimoto *et al.*, 2005), where it is simultaneously involved in epithelial morphogenesis and signaling mechanisms that drive microtubule repolymerization in developing epithelia. The two paralogs *RabX2* and *Rab9Fb* in *D. melanogaster* are more similar to the orthologs conserved across species and thus may have retained the ancestral function. The proteins encoded by *RabX2* and *Rab9Fb* are most similar to RAB1 protein, which is a key membrane traffic regulator. More precisely, RAB1 controls proteins transport between the ER and the Golgi apparatus (Wu *et al.*, 2003), by monitoring movement of G-protein–coupled receptors from the ER to the cell surface. When considering the implication of *Rab1* yeast ortholog in mitochondrial inheritance (Buvelot Frei *et al.*, 2006), it is interesting to note that among the 26 RAB proteins of *D. melanogaster*, RAB9Fb is the third RAB most similar to RAB1 yeast ortholog. It is noteworthy that RAB1 function is governed by GDI and RAB27 through biogenesis of melanosomes. As will be discussed below, it is tempting to postulate a possible link between *Gdi* and the *6paRAB9* because the resemblance between the phenotype of *D. melanogaster* larvae mutant for *Gdi* and the unrescued hybrids under study is striking.

V. GENERAL CONSIDERATIONS

Genetic studies based on large portions of genomes have supported the view that in *Drosophila* relatively few genes are involved in hybrid inviability, compared with genes involved in hybrid sterility (Coyne *et al.*, 1998; Hutter, 1987, 1997). Dobzhansky (1937) and Müller (1940) first pointed out that at least two genes must evolve independently in two isolated populations for a concomitant deleterious effect to result in HI. The findings discussed here suggest that inactivation of the *Rab9D* gene, which lies 1 kb from *Hmr*, has a deleterious effect in this species but reduces HI in hybrids between this species and its sibling species. *Rab9D* has two paralogs that are located about 32 and 38 kb away, whose nucleotide sequences are 98–99% identical to *Rab9D* and to each other.

Duplication of *6paRAB9* in *D. melanogaster* may have taken over part of *Hmr* function, rendering this gene partly redundant. The consecutive reduction in selective constraint on *Hmr* might have allowed it to evolve a novel function, involving the regulation of several genes coevolving with it in *D. melanogaster*. Several of these *Hmr*-regulated genes might have sufficiently diverged between species to result in impaired regulation by *D. melanogaster Hmr* in hybrid males. However, because the *6paRAB9* would have partially taken over the former function of *melanogaster Hmr*, a loss of function of this gene in hybrid males would not only suppress the above deleterious misregulation, but it would leave enough of the ancestral gene regulation as maintained by the *6paRAB9* to restore viability. In this view, duplication of the *6paRAB9* in *D. melanogaster* would have preceded divergence at the *Hmr-Lhr* genes, involving rapid evolution of at least eight genes in this barrier to reproduction.

Although results from interspecific crosses support an implication in HI of paralog *Rab9D*, the effects of inactivation of the other paralogs in hybrids have not been tested. Adding a wild-type copy of *Rab9D* gene in hybrids rescued by *Hmr1* does not suppress rescue, whereas a wild-type copy of *Hmr* clearly does so (Barbash *et al.*, 2003). This observation, however, may not rule out a role of HMR protein as a regulator of *Rab9D* and its paralogs. If HMR behaves as a positive effector of these paralogs, thus enhancing HI, adding a copy of only *Rab9D* may not suppress hybrid male rescue in the presence of the hypomorphic *Hmr1* allele. It should be borne in mind that a change in *Hmr* mRNA level was observed in only one (*Hmr1*) of the three hybrid-rescuing mutants that map to cytological region 9D-9E. Indeed, no causative mutation in *Hmr* has yet been characterized with respect to genomic DNA corresponding to the two remaining mutants *Df(1)EP307–1-2* and *In(1)AB* (Barbash *et al.*, 2003, 2004b). This fact, along with the observation of a marked decline in rescue observed in several *Hmr1* strains that retained the hybrid rescue causing P element, suggest that additional X-linked factors influence HI in hybrids between the species under study. In this regard, Presgraves (2003) and Wu and Ting (2004) reported evidence that multiple loci on the X chromosome of *D. melanogaster*, which interact with the *Nup96* gene in *D. simulans*, are involved in inviability of hybrids between these species.

Interspecific comparisons between *Rab9D* orthologs revealed a surprisingly high rate of AA replacement substitutions per site (D_n) over synonymous substitutions per site (D_s), which provides strong evidence that RAB9D function is a target of positive selection. The highest D_n/D_s ratio (2.63) was in the ancestor of the sibling species, an observation which is compatible with a causative role of *Rab9D* in HI between *D. melanogaster* and its sibling species. Thornton and Long (2002, 2005) analyzed the rates of divergence between 1841 pairs of genes duplicated in the genome of *D. melanogaster*. It is notable that these authors observed the highest K_a/K_s ratios between duplicates for *Rab9Fb*, *RabX2*, and *Rab9Db* genes. While two orthologs of the *6paRAB9* genes are active in the sibling species, only a single ortholog seems to exist in either *D. erecta* and or *D. yakuba*; and no ortholog is present in either *D. ananassae*, *D. pseudoobscura* or *D. persimilis*. The simplest evolutionary scenario to account for the increase in number of duplicated *Rab* genes would be that a parental copy evolved multifunctionality in the *melanogaster* subgroup, allowing for RAB subfunctionalization, with novel interactions becoming parceled out in duplicates. Such a scenario, as pioneered by Jensen (1976), could have resulted in formation of new links in biological networks, while old links would have been broken.

Barbash *et al.* (2004a) have shown that the *Hmr* gene is evolving rapidly, as revealed by many AA replacement substitutions, as well as several insertions/deletions between *D. melanogaster* and its sibling species. Although the amount of selection, as determined by the K_a/K_s ratio, is even greater at *Rab9D* than at *Hmr*, this may reflect important differences in protein structural constraints. Nevertheless, synteny analysis shows that *Hmr* and *Rab9D* have remained in tight association for about 10 Myr, with positive selection acting strongly on functions of both genes. This finding should encourage studies aimed at finding out whether *Hmr* and *Rab9D* are functionally linked. The rate of evolution at HMR was found to be particularly marked at its first domain that shares homology with MADF protein DNA-binding domain. In this regard, it is worth mentioning that the MADF domain recognizes the sequence TAACGG (Biedenkapp *et al.*, 1988), and that in all but one (*RabX2*) of the *6paRAB9* genes the translation-start codon is preceded by TAACG, a sequence that may represent a potential-binding site for a protein domain similar to MADF. If some interactions between *Hmr* and *Rab* paralogs were dependent on physical proximity to members of the *6paRAB9* cluster, this could be related to the rescue effect exerted by *Df(1)EP307–1-2* and *In(1)AB* alleles. The former mutation deletes about 61 kb of DNA within the *6paRAB9* cluster (Barbash *et al.*, 2004b), and the latter mutation splits the cluster, driving its *Rab* members apart (Hutter and Karch, 1994). The distal breakpoint of *In(1)AB* moves *Rab9Fa* and *Rab9Fb* genes from region 9 to region 13 of the X chromosome.

As revealed in the current article, the RAB9Fb and RABX2 proteins, which may have retained an ancestral function, are most similar to RAB1, whose function is governed by GDI (Rak *et al.*, 2003). GDI binds to GDP-bound RABs to remove these from acceptor membranes and to deliver them to donor membranes. In *D. melanogaster, Gdi* is present as a single gene, which is essential for development, and whose product binds to all the RABs from the genome. It is interesting that mutations in *Gdi* result in a phenotype (Ricard *et al.*, 2001) that closely resembles that of the poorly viable hybrid larvae—those that are not rescued by mutated *Hmr*. Three developmental processes are dependent on *Gdi* function, namely formation of pupal case, pole cell formation in the embryo, and signaling to imaginal discs. A critical defect in imaginal-disc development has been described in hybrids between *D. melanogaster* and its sibling species (Sanchez and Dübendorfer, 1983; Seiler and Nöthiger, 1974).

I have stressed that *Rab1* yeast ortholog plays a role in mitochondrial inheritance, an observation that lends support to a possible implication of *Rab9Fb* and *RabX2* genes in antagonistic coevolution of the *6paRAB9* between males and females, as argued before (Hutter, 2002). Since the *6paRab9* cluster spans about 550 kb of genomic DNA in *D. melanogaster*, it was speculated that additional genes from this cluster may influence HI, as a result of coevolution. Among these, two genes related to energy metabolism in mitochondria, *CG1628* and *sesB-Ant2*, which are involved in oxidative phosporylation (OXPHOS), have been discussed (Hutter, 2002; Hutter and Karch, 1994). Each of these OXPHOS genes was found to harbor a major alteration very near its untranslated regions in the two hybrid-rescuing mutants *Hmr1* and *In(1)AB*. The very large *sprint* gene (*CG 12638*), which lies only 2 kb from *Rab9D* in *D. melanogaster*, encodes a protein most similar to RAB5, which contains an RABEX nucleotide exchange factor for the small RAB5 GTPase (Szabo *et al.*, 2001). Again, it is intriguing that the *sprint* gene is partly deleted in the hybrid-rescuing mutant *Df(1)EP307–1-2* (Barbash and Ashburner, 2003). Since *Hmr* encodes a regulatory protein with homology to a family of MADF- and MYB-related DNA-binding transcriptional regulators, it is worth noting the functional link between MYB, GGA proteins, and RABEX-5 (Prag *et al.*, 2005). The Golgi-localized, γ-ear containing, Arf-binding (GGA) proteins are a family of adaptor proteins involved in vesicular transport. As an effector of ADP-ribosylation factor (ARF) GTPases, GGA interacts using its target of Myb (TOM) domain with GTP-bound form of ARF. The domain also interacts with Rabaptin-5, which is an effector of RAB5, as part of a complex with RABEX-5.

All small GTPases, including the RAB-related RAN GTPase and the molecular motor myosin, are thought to have evolved from a common ancestor, as they have been built around nucleotide-dependent molecular switches (Sablin and Fletterick, 2001). For instance, all 6paRAB9 proteins share a P-loop motif with ARF GTPases and myosin proteins. To achieve their functions,

RAB GTPases act along with specific motor proteins, such as actin-based myosin, to generate forces at the cytoskeleton, allowing the different steps of late endocytosis to take place. During this process, RAB GTPases can be regarded as central regulators of organelle dynamics (Goud, 2002), playing a pivotal role in the interaction and movement along the cytoskeleton. They participate in the intense traffic of proteins in and out of the nucleus, which is mediated by a series of finely choreographed protein–protein interactions, including those that connect the junctional complex to specific elements of the cytoskeleton. In this regard, it is interesting that three of the *6paRab9* genes located at both edges of the cluster are nested within large genes related to motor proteins (Fig. 1.1). *RabX2* lies in the seventh intron of the *CG 32683* gene, which is involved in intracellular protein transport; and *Rab9Fa* and *Rab9Fb* are nested within the large *Myosin10A* gene, which is involved in cytoskeleton organization and biogenesis.

Another gene of particular interest lies just outside the *6paRab9* cluster in *D. melanogaster*'s cytological band 10A8 and encodes the RAN small GTPase. This protein, which shares a domain with RAB9D, is the only source of energy required for nucleocytoplasmic exchange, and it controls the transport of molecules across the nuclear pore complex. A direct role of components of the RAN/Importin systems in mitotic spindle has been established, whereby activation of RAB proteins enables to recruit various effector proteins, which influence not only vesicular transport but also vesicle movement along cytoskeletal filaments (Salina *et al.*, 2003). Proper control of cell division thus relies on the faithful and coordinated actions of proteins involved in nuclear pore complex assembly and nucleocytoplasmic transport. In *D. simulans*, the *Nup96* gene, which encodes a subunit of the nuclear pore complex transporting macromolecules between the nucleus and the cytoplasm, is strongly involved in hybrid inviability (Presgraves *et al.*, 2003).

It should be emphasized that inactivation of the *Rab9D* gene in a hybrid genotype appears to be associated with an absence of otherwise typical necrotic tissue patches, which points to a possible role of this gene in early tumorigenesis in hybrids. In mammals the role of RAB proteins in tumor formation is well established, as illustrated by RNA microarray analyses; they show that nearly 50% of known *Rab* genes display increases in mRNA expression in human ovarian cancer (Cheng *et al.*, 2005). The similarity between RAB9D and RAB27, besides the above-mentioned link between RAB and myosin proteins, suggests a possible role of the *6paRab9* cluster in melanosomes biogenesis. Melanosome transport is mainly regulated by actin-based myosin proteins, and RAB27a is required to recruit myosin Va to melanosomes, thereby tethering melanosomes to the peripheral actin network (Hume *et al.*, 2002). Interestingly, the gene defective in *leaden* mice encodes melanophilin, which has strong homology with RAB3a effectors. In *D. melanogaster* the gene coding for rabphilin 3A is located in the middle of the *6paRAB9* cluster (Fig. 1.1), and its product is known to establish a specific interaction between each RAB protein and its effectors (Merithew *et al.*, 2001). While addressing the possible role of melanocytes in HI, it is worth noting that one of the four HI genes now characterized at the molecular level is the melanoma-inducing *Tu* locus in *Xiphophorus* (Wittbrodt *et al.*, 1989). In their analysis of the *Xmrk-2* gene, the authors were able to implicate the MAP kinase cascade in intracellular signaling in pigment cells and cell transformation in hybrids. Thus, *Xmrk-2* appears to be misexpressed in hybrids, causing tumor formation (Malitschek *et al.*, 1995). An *Xmrk-1* gene was found to be adjacent to *Xmrk-2*, and the latter probably originated from nonhomologous recombination between *Xmrk-1* and a linked locus (see in Wu and Ting, 2004).

A prediction from an "HI model" proposed previously (Hutter, 2002) is that antagonistic coevolution between males and females should favor rapid divergence of genes of the *6paRAB9* cluster. This is in line with the major picture emerging from genetics of speciation, that is, that genes causing post-zygotic isolation are likely to be rapidly evolving and to diverge by positive Darwinian selection (Coyne and Orr, 2004). This signature of speciation genes is clearly visible in proteins encoded by *Hmr* (Barbash *et al.*, 2003, 2004a), by *Nup96* (Presgraves, 2003), and by the *6paRAB9* genes discussed here. Further support of a fast rate of evolution of speciation genes in *Drosophila* comes from studies of the *Odysseus* gene, that influences HI by causing hybrid sterility. This gene was not only shown to evolve rapidly, but also in relation to gene duplication events (Ting *et al.*, 2004; Wu and Ting, 2004). Remarkably, results from expression analysis, transgenic experiments, and knockout of *Odysseus* suggested that the HI phenotype represents a novel manifestation of the gene function in hybrids (Sun *et al.*, 2004). Similarly, studying expression of the *6paRab9* genes in *D. melanogaster*, and of their orthologs in the sibling species, is likely to shed light on the evolution of species isolation in *Drosophila*.

References

Barbash, D. A., and Ashburner, M. (2003). A novel system of fertility rescue in *Drosophila* hybrids reveals a link between hybrid lethality and female sterility. *Genetics* **163,** 217–226.

Barbash, D. A., Roote, J., and Ashburner, M. (2000). The *Drosophila melanogaster* hybrid male rescue gene causes inviability in male and female species hybrids. *Genetics* **154,** 1747–1771.

Barbash, D. A., Siino, D. F., Tarone, A. M., and Roote, J. (2003). A rapidly evolving MYB-related protein causes species isolation in *Drosophila*. *Proc. Natl. Acad. Sci. USA* **100,** 5302–5307.

Barbash, D. A., Awadalla, P., and Tarone, A. M. (2004a). Functional divergence caused by ancient positive selection of a *Drosophila* hybrid incompatibility locus. *PLoS Biol.* **2**(6), e142.

Barbash, D. A., Roote, J., Johnson, G., and Ashburner, M. (2004b). A new hybrid rescue allele in *Drosophila melanogaster*. *Genetica* **120,** 261–266.

Bergman, C. M., Pfeiffer, B. D., Rincon-Limas, D. E., Hoskins, R. A., Gnirke, A., Mungall, C. J., Wang, A. M., Kronmiller, B., Pacleb, J., Park, S., Stapleton, M., Wan, K., *et al.* (2002). Assessing the impact of comparative genomic sequence data on the functional annotation of the *Drosophila* genome. *Genome Biol.* **3**, 1–20.

Biedenkapp, H., Borgmeyer, U., Sippel, A. E., and Klempnauer, K. H. (1988). Viral myb oncogene encodes a sequence-specific DNA-binding activity. *Nature* **335**, 835–837.

Buvelot Frei, S., Rahl, P. B., Nussbaum, M., Briggs, B. J., Calero, M., Janeczko, S., Regan, A. D., Chen, C. Z., Barral, Y., Whittake, G. R., and Collins, R. N. (2006). Bioinformatic and comparative localization of Rab proteins reveals functional insights into the uncharacterized GTPases Ypt10p and Ypt11p. *Mol. Cell. Biol.* **26**, 7299–7317.

Charlesworth, B., Coyne, J. A., and Barton, N. H. (1987). The relative rates of evolution of sex chromosomes and autosomes. *Am. Nat.* **130**, 113–146.

Cheng, K. W., Lahad, J. P., Gray, J. W., and Mills, G. B. (2005). Emerging role of RAB GTPases in cancer and human disease. *Cancer Res.* **65**, 2516–2519.

Cordon, M. (1927). The genetics of a viviparous top-minnow *Platypoecilus:* The inheritance of two kinds of melanophores. *Genetics* **12**, 253–283.

Coyne, J. A., and Orr, H. A. (2004). "Speciation." Sinauer Associates Inc., Sunderland, MA.

Coyne, J. A., Simeonidis, S., and Rooney, P. (1998). Relative paucity of genes causing inviability in hybrids between *Drosophila melanogaster* and *Drosophila simulans*. *Genetics* **150**, 1091–1103.

David, J., Lemeunier, F., Tsacas, L., and Bocquet, C. (1974). Hybridization of a new species, *Drosophila mauritiana*, with *D. melanogaster* and *D. simulans*. *Ann. Genet.* **17**, 235–241.

David, J., Bocquet, C., and Pla, E. (1976). New results on the genetic characteristics of the Far East race of *Drosophila melanogaster*. *Genet. Res.* **28**, 253–260.

Dobzhansky, T. (1937). "Genetics and the Origin of Species." Columbia University Press, New York.

Goud, B. (2002). How Rab proteins link motors to membranes. *Nat. Cell Biol.* **4**, E77–E78.

Hadorn, E. (1961). Zur Autonomie und Phasenspezifität der Letalität von Bastarden zwischen *Drosophila melanogaster* und *Drosophila simulans*. *Rev. Suisse Zool.* **68**, 197–207.

Hume, A. N., Collinson, L. M., Hopkins, C. R., Strom, M., Barral, D. C., Bossi, G., Griffiths, G. M., and Seabra, M. C. (2002). The leaden gene product is required with Rab27a to recruit myosin Va to melanosomes in melanocytes. *Traffic* **3**, 193–202.

Hutter, P. (1987). Pleiotropic effects of environment-sensitive genes affecting fitness in relation to postmating reproductive isolation. *Genetica* **72**, 193–198.

Hutter, P. (1997). Genetics of Hybrid Inviability in *Drosophila*. *Adv. Genet.* **36**, 157–185.

Hutter, P. (2002). X-linked small GTPase and OXPHOS genes are candidates for the genetic basis of hybrid inviability in *Drosophila*. *Dev. Genes Evol.* **212**, 504–512.

Hutter, P., and Ashburner, M. (1987). Genetic rescue of inviable hybrids between *Drosophila melanogaster* and its sibling species. *Nature* **327**, 331–333.

Hutter, P., and Karch, F. (1994). Molecular analysis of a candidate gene for the reproductive isolation between sibling species of *Drosophila*. *Experientia* **50**, 249–262.

Hutter, P., Roote, J., and Ashburner, M. (1990). A genetic basis for the inviability of hybrids between sibling species of *Drosophila*. *Genetics* **124**, 909–920.

Jensen, R. A. (1976). Enzyme recruitment in evolution of new function. *Annu. Rev. Microbiol.* **30**, 409–425. Review.

Kosswig, C. (1927). Uber Bastarde der Teteostier *Platypoecilus* und Xiphophorus. *Zeitschrift für induktive Abstammungs und Vererbungslehre* **44**, 253–255.

Lachaise, D., Cariou, M. L., David, J. R., Lemeunier, F., Tsacas, L., and Ashburner, M. (1988). Historical biogeography of the *Drosophila melanogaster* species subgroup. *Evol. Biol.* **22**, 159–225.

Malitschek, B., Fornzler, D., and Schartl, M. (1995). Melanoma formation in Xiphophorus: A model system for the role of receptor tyrosine kinases in tumorigenesis. *Bioessays* **17**, 1017–1023. Review.

Mattaj, I. W., and Englmeier, L. (1998). Nucleocytoplasmic transport: The soluble phase. *Annu. Rev. Biochem.* **67,** 265–306.

Mayr, E. (1943). "Sytematics and the Origin of Species." Columbia University Press, New York.

Merithew, E., Hatherly, S., Dumas, J. J., Lawe, D. C., Heller-Harrison, R., and Lambright, D. G. (2001). Structural plasticity of an invariant hydrophobic triad in the switch regions of Rab GTPases is a determinant of effector recognition. *J. Biol. Chem.* **276,** 13982–13988.

Morimoto, S., Nishimura, N., Terai, T., Manabe, S., Yamamoto, Y., Shinahara, W., Miyake, H., Tashiro, S., Shimada, M., and Sasaki, T. (2005). Rab13 mediates the continuous endocytic recycling of occludin to the cell surface. *J. Biol. Chem.* **280,** 2220–2228.

Müller, H. J. (1940). Bearings of the *Drosophila* work on systematics. *In* "The New Systematics" (J. S. Huxley, ed.), pp. 185–268. Oxford University Press (Clarendon), Oxford.

Musch, A. (2004). Microtubule organization and function in epithelial cells. *Traffic* **5,** 1–9.

Nei, M., and Gojobori, T. (1986). Simple methods for estimating the numbers of synonymous and nonsynonymous nucleotide substitutions. *Mol. Biol. Evol.* **3,** 418–426.

Pfeffer, S. R. (2001). Rab GTPases: Specifying and deciphering organelle identity and function. *Trends Cell Biol.* **11,** 487–491.

Powell, J. R. (1997). "Progress and Prospects in Evolutionary Biology: The *Drosophila* Model," p. 580. Oxford University Press, New York.

Prag, G., Lee, S., Mattera, R., Arighi, C. N., Beach, B. M., Bonifacino, J. S., and Hurley, J. H. (2005). Structural mechanism for ubiquitinated-cargo recognition by the Golgi-localized, gamma-ear-containing, ADP-ribosylation-factor-binding proteins. *Proc. Natl. Acad. Sci. USA* **102,** 2334–2339.

Presgraves, D. C. (2003). A fine-scale genetic analysis of hybrid incompatibilities in *Drosophila*. *Genetics* **163,** 955–972.

Presgraves, D. C., Balagopalan, L., Abmayr, S. M., and Orr, H. A. (2003). Adaptive evolution drives divergence of a hybrid inviability gene between two species of *Drosophila*. *Nature* **423,** 715–719.

Rak, A., Pylypenko, O., Durek, T., Watzke, A., Kushnir, S., Brunsveld, L., Waldmann, H., Goody, R. S., and Alexandrov, K. (2003). Structure of Rab GDP-dissociation inhibitor in complex with prenylated YPT1 GTPase. *Science* **302,** 646–650.

Ricard, C. S., Jakubowski, J. M., Verbsky, J. W., Barbieri, M. A., Lewis, W. M., Fernandez, G. E., Vogel, M., Tsou, C., Prasad, V., Stahl, P. D., Waksman, G., and Cheney, C. M. (2001). *Drosophila* rab GDI mutants disrupt development but have normal Rab membrane extraction. *Genesis* **31,** 17–29.

Riggs, B., Rothwell, W., Mische, S., Hickson, G. R., Matheson, J., Hays, T. S., Gould, G. W., and Sullivan, W. (2003). Actin cytoskeleton remodeling during early *Drosophila* furrow formation requires recycling endosomal components nuclear-fallout and Rab11. *J. Cell Biol.* **163,** 143–154.

Sablin, E. P., and Fletterick, R. J. (2001). Nucleotide switches in molecular motors: Structural analysis of kinesins and myosins. *Curr. Opin. Struct. Biol.* **11,** 716–724.

Salina, D., Enarson, P., Rattner, J. B., and Burke, B. (2003). Nup358 integrates nuclear envelope breakdown with kinetochore assembly. *J. Cell Biol.* **162,** 991–1001.

Sanchez, L., and Dübendorfer, A. (1983). Development of imaginal discs from lethal hybrids between *Drosophila melanogaster* and *Drosophila mauritiana*. Wihelm Roux's. *Arch. Dev. Biol.* **192,** 48–50.

Sawamura, K., Taira, T., and Watanabe, T. K. (1993a). Hybrid lethal systems in the *Drosophila melanogaster* species complex. I. The *maternal hybrid rescue* (*mhr*) gene of *Drosophila simulans*. *Genetics* **133,** 299–305.

Sawamura, K., Yamamoto, M. T., and Watanabe, T. K. (1993b). Hybrid lethal systems in the *Drosophila melanogaster* species complex. II. The *Zygotic hybrid rescue* (*Zhr*) gene of *D. melanogaster*. *Genetics* **133,** 307–313.

Schuler, G. D., Altschul, S. F., and Lipman, D. J. (1991). A workbench for multiple alignment construction and analysis. *Proteins Struct. Funct. Genet.* **9,** 180–190.

Seabra, M. C., Mules, E. H., and Hume, A. N. (2002). Rab GTPases, intracellular traffic and disease. *Trends Mol. Med.* **8,** 23–30.

Seiler, T., and Nöthiger, R. (1974). Somatic cell genetics applied to species hybrids of *Drosophila*. *Experientia* **30,** 709(abstract).

Sturtevant, A. H. (1920). Genetic studies on *Drosophila simulans*. I. Introduction. Hybrids with *Drosophila melanogaster*. *Genetics* **5,** 488–500.

Sun, S., Ting, C. T., and Wu, C. I. (2004). The normal function of a speciation gene, *Odysseus*, and its hybrid sterility effect. *Science* **305,** 81–83.

Szabo, K., Jekely, G., and Rorth, P. (2001). Cloning and expression of *sprint*, a *Drosophila* homologue of *RIN1*. *Mech. Dev.* **101,** 259–262.

Thibault, S. T., Singer, M. A., Miyazaki, W. Y., Milash, B., Dompe, N. A., Singh, C. M., Buchholz, R., Demsky, M., Fawcett, R., Francis-Lang, H. L., Ryner, L., Cheung, L. M., *et al.* (2004). A complementary transposon tool kit for *Drosophila melanogaster* using P and piggyBac. *Nat. Genet.* **36,** 283–287.

Thornton, K., and Long, M. (2002). Rapid divergence of gene duplicates on the *Drosophila melanogaster* X chromosome. *Mol. Biol. Evol.* **19,** 918–925.

Thornton, K., and Long, M. (2005). Excess of amino acid substitutions relative to polymorphism between X-linked duplications in *Drosophila melanogaster*. *Mol. Biol. Evol.* **22,** 273–284.

Ting, C. T., Tsaur, S. C., Sun, S., Browne, W. E., Chen, Y. C., Patel, N. H., and Wu, C. I. (2004). Gene duplication and speciation in *Drosophila*: Evidence from the Odysseus locus. *Proc. Natl. Acad. Sci. USA* **101,** 1232–1235.

Watanabe, T. K. (1979). A gene that rescues the lethal hybrids between *Drosophila melanogaster* and *D. simulans*. *Jpn. J. Genet.* **54,** 325–331.

Wittbrodt, J., Adam, D., Malitschek, B., Maueler, W., Raulf, F., Telling, A., Robertson, S. M., and Schartl, M. (1989). Novel putative receptor tyrosine kinase encoded by the melanoma-inducing Tu locus in *Xiphophorus*. *Nature* **341,** 415–421.

Wu, C. I., and Ting, C. T. (2004). Genes and speciation. *Nat. Rev. Genet.* **5,** 114–122.

Wu, G., Zhao, G., and He, Y. (2003). Distinct pathways for the trafficking of angiotensin II and adrenergic receptors from the endoplasmic reticulum to the cell surface: Rab1-independent transport of a G protein-coupled receptor. *J. Biol. Chem.* **278,** 47062–47069.

Yang, Z. (2004). PAML: Phylogenetic analysis by maximum likelihood. Version 3.14, University College London, London.

2

The *Neurospora crassa* Circadian Clock

Christian Heintzen* and Yi Liu†

*Faculty of Life Sciences, University of Manchester, Manchester M13 9PT
United Kingdom
†Department of Physiology, University of Texas Southwestern Medical Center
Dallas, Texas 75390

Advances in Genetics, Vol. 58
Copyright 2007, Elsevier Inc. All rights reserved.

0065-2660/07 $35.00
DOI: 10.1016/S0065-2660(06)58002-2

ABSTRACT

The filamentous fungus *Neurospora crassa* is one of a handful of model organisms that has proven tractable for dissecting the molecular basis of a eukaryotic circadian clock. Work on Neurospora and other eukaryotic and prokaryotic organisms has revealed that a limited set of clock genes and clock proteins are required for generating robust circadian rhythmicity. This molecular clockwork is tuned to the daily rhythms in the environment via light- and temperature-sensitive pathways that adjust its periodicity and phase. The circadian clockwork in turn transduces temporal information to a large number of clock-controlled genes that ultimately control circadian rhythms in physiology and behavior. In summarizing our current understanding of the molecular basis of the Neurospora circadian system, this chapter aims to elucidate the basic building blocks of model eukaryotic clocks as we understand them today.   © 2007, Elsevier Inc.

I. INTRODUCTION

Circadian clocks are cellular timekeepers that impart temporal organization on a large number of molecular, physiological, and behavioral activities and are found in most eukaryotic and some prokaryotic life forms (Dunlap *et al.*, 2004; Young and Kay, 2001). The intrinsic periodicity of circadian clocks, which is roughly but not exactly 24 h, suggests that they have evolved under the selective forces imposed by the daily rhythms in the Earth's environment. Thus, circadian clocks are an adaptive response that facilitate optimal survival of organisms, for example, by allowing them to anticipate the predictable environmental rhythms that occur on our planet. Although circadian clocks are not essential for an organism's survival, there is evidence that at the population level, in a rhythmically changing environment and competitive situation, circadian clocks can be critical for survival (Ouyang *et al.*, 1998).

Circadian clocks are autonomous timers, that is they measure time by self-excitatory cyclical processes that enable them to preserve information on time in the absence of environmental cues. However, in the natural environment, they generally perform their main adaptive function in a rhythmically changing environment. Under these conditions, circadian clocks are coupled to the 24-h periodicity of the external world in a process called entrainment

(Johnson *et al.*, 2003; Roenneberg *et al.*, 2003). A key feature of entrained circadian clocks (and in contrast to passively driven or masked phenomena) is their ability to impose phase information on rhythmic activities, that is they actively influence the time of day when things happen in the cell. Not surprisingly, light and temperature are the most important cues for the entrainment of circadian clocks to the outside world.

Light and temperature can act acutely to perturb and thus reset circadian clocks in a universally predictable manner: a perturbing pulse will delay circadian clocks when given in the late subjective day/early subjective night, advance them when given during the late subjective night, and elicit only small responses when given during the subjective day. This universal behavior helps circadian clocks to align their day and night phases with that of the environment (Johnson *et al.*, 2003). In contrast to these acute effects, the chronic effects of light and temperature on the circadian system are different. For example, the free-running period of circadian clocks is relatively unaffected across a wide range of physiologically relevant temperatures, that is the clock is temperature compensated. In contrast, chronic light treatment often results in arrhythmia or, depending on the intensity of light, can influence the period length of the clock in a mechanism that may play a role in the entrainment process to full photoperiods (Johnson *et al.*, 2004). Taken together, a number of defining features have been identified that distinguish circadian clocks from a number of other rhythmic but noncircadian phenomena.

This chapter will summarize the progress that has been made in recent years in understanding the molecular basis of circadian rhythmicity in the simple eukaryote *Neurospora crassa*.

II. *NEUROSPORA* AS A CIRCADIAN MODEL SYSTEM

Neurospora was introduced to circadian clock research almost 50 years ago and detailed accounts of how it became an established clock model have been given before and shall not be repeated here (Dunlap and Loros, 2004; Loros and Dunlap, 2001). Instead, we will give only a very brief introduction of Neurospora as a circadian model system, before providing a more detailed treatment of the molecular basis of rhythmicity in this fungus.

Neurospora's utility for circadian clock research arises chiefly from a combination of a self-recording circadian phenotype and a set of well-developed molecular and genetic tools. A fully sequenced genome as well as improvements and additions to the molecular tool set have further boosted Neurospora's value as a model organism for the study of circadian and other problems in biology (Borkovich *et al.*, 2004; Colot *et al.*, in press; Galagan *et al.*, 2003). The most commonly used method by which the Neurospora circadian system is visualized employs the so-called race tube assay. Race tubes are glass tubes filled with a growth medium on which the fungus is inoculated (Loros and Dunlap, 2001). While "racing" down the tube, Neurospora records its own circadian rhythmicity by producing zones of sparse mycelia that are followed by denser "bands" of asexual spore producing aerial hyphae. This self-recording behavior is a unique and convenient feature of the Neurospora circadian system and repeats approximately every 22 h for a wild-type Neurospora strain kept in constant darkness (DD) and constant temperature. The characteristics of this rhythmically controlled event in development are circadian, that is the observed rhythm in asexual spore formation (conidiation) free-runs with approximately daily periodicity, can be phase-shifted and entrained by light and temperature cues, and is temperature and nutritionally compensated. An exhaustive treatment of the race tube assay, its merits, and caveats has been given elsewhere (Dunlap and Loros, 2004, 2005; Gooch et al., 2004). Other methods by which the state of the circadian clock can be assessed involve the analysis of molecular phenotypes such as monitoring rhythms in levels of clock gene products by standard molecular techniques. A firefly luciferase reporter system has been developed for Neurospora in which a clock-controlled promoter fused to the luciferase gene can faithfully record circadian activity at the ccg-2 promoter (Morgan et al., 2003). However, it was the ease of the race tube assay that has played the main role in the identification of clock mutants and the discovery (Feldman and Hoyle, 1973) and eventual cloning (McClung et al., 1989) of the *frequency* (*frq*) locus, the first central circadian clock element identified in Neurospora. Following the successful cloning of the *frq* gene and the appreciation of its importance for circadian rhythmicity, the ensuing molecular work has centered around identifying components that are functionally linked to the *frq* gene. This chapter will concentrate on what we have learned about the molecular components and regulations of this circadian oscillator, its interaction with the rhythmic environment, and its control of target genes and overt rhythmicity. It will also highlight some of the more recent work on oscillators that operate in the absence of the *frq* gene, commonly referred to as FRQ-less oscillators (FLOs).

III. THE MOLECULAR MACHINERY OF THE NEUROSPORA CIRCADIAN OSCILLATOR

One central goal of circadian clock studies is to understand how a circadian oscillator is assembled and regulated at the molecular level to generate endogenous circadian rhythmicity. Identification of clock components and understanding their regulation and interactions are essential to achieve this. Since the positional cloning of *frq*, the second eukaryotic circadian clock gene to be cloned (McClung et al., 1989), more than 10 components of the Neurospora circadian oscillator have been identified by genetic (both forward and reverse) and biochemical approaches (Table 2.1). Among these, FRQ, FRH, WC-1, and WC-2 can be considered to lie at the core of the Neurospora circadian oscillator. The remaining genes and their products have roles in posttranslational regulation and degradation of FRQ, underlining the importance of these modifications in the regulation of the Neurospora circadian negative feedback loop. The identification of these clock components and studies of their functions have

Table 2.1. Components in the Neurospora Circadian Oscillator

Clock proteins	Molecular functions in the clock
FRQ	Essential for circadian clock function. Forms a complex with FRH and acts as the negative element in the circadian negative feedback loop by repressing WCC activity.
FRH	Essential for circadian clock function. Forms a complex with FRQ and acts as the negative element by repressing WCC activity. An essential gene in Neurospora.
WC-1	Essential for circadian clock function. Forms a heterodimeric complex with WC-2 and acts as the positive element in the circadian negative feedback loop by activating *frq* transcription. It is also the blue-light photoreceptor required for the light resetting of the clock and all other known light responses in Neurospora.
WC-2	Essential for circadian clock function. Forms a heterodimeric complex with WC-1 and acts as the positive element in the circadian negative feedback loop by activating *frq* transcription. It is also required for the light resetting of the clock and all other known light responses in Neurospora.
Casein kinase 1a	An important and conserved clock component that binds and phosphorylates FRQ to promote its degradation mediates FRQ-dependent WC phosphorylation to close the circadian negative feedback loop.
Casein kinase II (CKA, CKB1, and CKB2)	Essential for circadian clock function. Phosphorylates FRQ *in vivo* and *in vitro*. Its phosphorylation of FRQ regulates FRQ stability and is important for the repressor activity of FRQ. Also mediates FRQ-dependent WC phosphorylation.
CAMK-1	A major kinase that phosphorylates FRQ *in vitro*. Disruption leads to changes in light-induced phase-shifts and modest change of period. Two additional homologues in Neurospora.
PP1	Regulates the stability of FRQ probably by dephosphorylating FRQ.
PP2A (RGB-1, the regulatory subunit)	Important for the closing of the negative feedback loop, probably by dephosphorylating FRQ and WC proteins.
FWD-1	Essential for circadian clock function. Interacts with FRQ and functions as the substrate recruiting subunit of an SCF-type ubiquitin ligase that mediates FRQ ubiquitination and degradation.
CSN (COP9 signalosome)	Important for normal clock functions by regulating the stability of the SCF^{FWD-1} complex.

made the Neurospora circadian oscillator one of the best understood eukaryotic circadian clock systems.

A. The Neurospora circadian feedback loops

Similar to the circadian oscillators in Drosophila and mammals, at the core of the Neurospora circadian oscillator lies an autoregulatory negative feedback loop, that is essential for the generation of circadian rhythmicity (Dunlap, 1999; Young and Kay, 2001). Within this and other eukaryotic circadian negative feedback loops, three processes can be distinguished that are essential for maintaining circadian rhythmicity. First, the activation of negative element(s) by positive elements so that the levels of the negative element increase. Second, the inhibition of the activation process by the negative elements, which prevents further increase of the negative element. Third, the release of inhibition and restoration of activation, achieved by the proper degradation of the negative element and activation of positive elements, which subsequently allows the reactivation of the negative element at the appropriate time to restart a new cycle. Each of these processes must be properly controlled to generate circadian rhythms that are close to 24 h.

In Neurospora, WC-1 and WC-2, the two PER-ARNT-SIM (PAS) domain-containing transcription factors, are the positive elements whereas FRQ, in complex with the RNA helicase, FRH, functions in the negative limb of the circadian feedback loop (Aronson et al., 1994a; Cheng et al., 2005; Crosthwaite et al., 1997; Linden et al., 1999). Figure 2.1 depicts our current understanding of the Neurospora circadian feedback loops. During late subjective night in constant darkness, a nuclear heterodimeric WC complex (D-WCC) consisting of WC-1 and WC-2 activates *frq* transcriptionally by binding to its promoter (Froehlich et al., 2003). Transcriptional activation leads to an increase of *frq* mRNA levels (Cheng et al., 2001b; Crosthwaite et al., 1997), which peak in the early subjective morning (Aronson et al., 1994a). Subsequently, FRQ protein levels peak in the late subjective day, that is with a 4- to 6-h delay relative to the *frq* mRNA oscillation (Garceau et al., 1997). At the same time, FRQ dimerizes with itself and forms a complex with FRH (FFC, FRQ–FRH complex) (Cheng et al., 2001a, 2005).

After its entry into the nucleus, FFC represses D-WCC activity, resulting in a decrease of *frq* mRNA levels beginning around mid-subjective day and reaching its low point around mid-subjective night (Aronson et al., 1994a). The inhibition of white collar complex (WCC) activity is likely mediated by the FFC–WCC interaction (Cheng et al., 2001a, 2005; Denault et al., 2001; Merrow et al., 2001), which may lead to the phosphorylation of WC proteins (Schafmeier et al., 2005). Phosphorylation of WC proteins results in the decrease of their DNA-binding activity and decrease of *frq* transcription

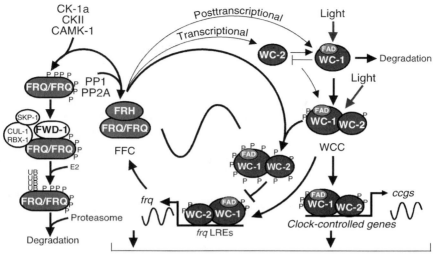

Figure 2.1. Schematic drawing of the core circadian feedback loops and the associated degradation pathways as far as they are known. The central circadian oscillator is a transcription–translation-based feedback circuit in which a hypophosphorylated WC-1:WC-2 heterodimeric complex (WCC) binds to the frq promoter and activates frq transcription. Following its translation, FRQ dimerizes and forms a complex with an FRQ-directed RNA helicase (FRH) called the FFC. The FFC mediates WCC hyperphosphorylation, possibly by recruiting a kinase to the WCC, thus rendering the WCC transcriptionally inactive and closing the negative feedback loop. The fact that repressor activity of FFC requires proper FRQ phosphorylation by CKII has been omitted from this cartoon for reasons of simplicity. At the transcriptional and posttranscriptional level, FFC stimulates the production of WC-1 and WC-2. Similarly, WC-1 and WC-2 negatively and positively regulate each others levels. The mechanisms of these regulations are unknown but these additional regulatory loops are thought to make the central oscillator more robust. Light can enter the system via the WC-1 protein (or the WCC) that binds the chromophore FAD and acts as the main blue-light photoreceptor in Neurospora. The WCC also binds to downstream targets of the circadian oscillator and thus may confer rhythmicity and light responses of clock-controlled output genes, which ultimately help to control overt circadian rhythmicity. Timely degradation of the FFC is critical for maintaining robust rhythmicity and is controlled at the level of FRQ phosphorylation by the kinases CK-1a, CKII, and CAMK-1 and phosphatases PP1 and PP2A. Phosphorylation targets FRQ for degradation and is mediated via the ubiquitin/proteasome. FRQ binds transiently to FWD-1, an F-Box/WD40 repeat-containing adaptor that serves as the substrate recruiting subunit of the SCF-type ubiquitin (E3) ligases. Other components of the SCF complex are SKP-1, Cullin, and RBX-1, a RING domain-containing protein. On binding to the SCF complex, FRQ becomes quickly ubiquitinated and degraded by the proteasome. The stability of the SCF complex is regulated by the COP9 signalosome (not shown), adding another layer of regulation to FRQ degradation. For more detail see text. (See Color Insert.)

(He and Liu, 2005; He et al., 2005b; Schafmeier et al., 2005). In addition, the repressive activity of FFC also requires proper phosphorylation of FRQ proteins by casein kinase II (CKII) (Yang et al., 2002, 2003, 2004).

To release the repression of D-WCC by FFC, FRQ protein needs to be properly degraded since the level of FFC is determined by the amount of FRQ (Cheng et al., 2005). FRQ is degraded through a phosphorylation-dependent ubiquitination/proteasome-mediated pathway (Gorl et al., 2001; He et al., 2003; Liu, 2005; Liu et al., 2000; Yang et al., 2001, 2002, 2003, 2004). As soon as FRQ is synthesized, it is immediately and progressively phosphorylated over time by several kinases, including casein kinase 1a (CK-1a), CKII, and a calcium/calmodulin (Ca/CaM)-dependent kinase (calmodulin kinase 1, CAMK-1). After FRQ becomes extensively phosphorylated, it physically interacts with FWD-1, an F-box/WD-40 repeat-containing protein and the substrate-recruiting subunit of an SCF-type ubiquitin ligase complex, which mediates the ubiquitination of FRQ and its degradation by the proteasome system. When the levels of FRQ fall below a critical threshold, the repression of D-WCC by FFC is released and *frq* transcription is reactivated by WCC to start a new cycle at late subjective night.

Aside from repressing WCC activity, FRQ also positively regulates the expression of both WC-1 and WC-2, thus forming positive feedback loops that are interlocked with the negative feedback loop (Cheng et al., 2001b, 2002; Lee et al., 2000; Merrow et al., 2001). The underlying mechanisms of such FRQ actions are not clear, but it is known that FRQ regulates WC-1 expression posttranscriptionally independent of WC-2 (Cheng et al., 2001b, 2002; Lee et al., 2000), while it promotes WC-2 expression at the level of transcription (Cheng et al., 2001b). Similar interlocked positive feedback loops have been found in Drosophila and mammals, suggesting that they may comprise a common regulatory mechanism among eukaryotic circadian oscillators (Glossop et al., 1999; Shearman et al., 2000).

By constitutively expressing WC-1 or WC-2, it was shown that these positive feedback loops are important for the robustness and stability of the Neurospora circadian clock since both the physiological and molecular rhythms are more pronounced when levels of WC proteins are artificially increased (Cheng et al., 2001b). Interestingly, although the robustness of the circadian conidiation rhythms could be changed in these strains by changing the levels of constitutively expressed WC-1 and WC-2 (and FRQ), the period of circadian conidiation varied little, a phenomenon later also observed in Drosophila (Kim et al., 2002). This observation suggests that, whereas rhythmicity of the WCs is not important for a functional feedback loop, the interlocked nature of the circadian feedback loops may be important for setting the period of the clock to be around 24 h, a property perhaps important for temperature compensation of the clock. Similar conclusions were also drawn from studies in Drosophila and mammalian systems (Kim et al., 2002; Preitner et al., 2002).

WC-1 and WC-2 also regulate each other to form another interacting feedback loop: WC-2 positively regulates WC-1 level by forming the WC complex, while WC-1 negatively regulates transcription of *wc-2* at the transcriptional level (Cheng et al., 2002, 2003b). Perhaps, the feedback regulation among the *wc* genes maintains levels of WCC that are both optimal for carrying out its function in the circadian clock and blue-light sensing and signaling.

B. WC-1 and WC-2: The activators of *frq* transcription in the circadian negative feedback loop

Like the positive elements in the Drosophila and mammalian oscillators, both WC-1 and WC-2 are PAS domain transcription factors (Ballario et al., 1996; Linden and Macino, 1997). WC-1 has three PAS domains and WC-2 has only one, and both proteins are GATA-type zinc (Zn)-finger proteins that are predominantly nuclear localized (Lee et al., 2000; Schwerdtfeger and Linden, 2000). The most N-terminal PAS domain of WC-1 is a specialized domain also known as light-, oxygen-, or voltage-sensing (LOV) domain and functions as the blue-light sensory module that mediates all known blue-light responses in Neurospora (Liu, 2003; Liu et al., 2003), while its other less-conserved PAS domains (PASB and PASC) show some similarity to those in the mammalian and Drosophila clock proteins BMAL1 and CYC (Lee et al., 2000).

WC-1 and WC-2 form heteromeric complexes *in vivo* mediated by the PASC domain of WC-1 and the single PAS domain of WC-2 (Cheng et al., 2002, 2003b; Talora et al., 1999). Levels of WC-1 protein are extremely low in the *wc-2* knockout strain or in strains in which the WC-1–WC-2 interaction is disrupted, indicating that WC-2 is required for maintaining WC-1 levels through formation of the WCC (Cheng et al., 2002). It is likely that WC-1 is unstable or cannot fold properly by itself.

In *wc* mutants, both *frq* mRNA and FRQ protein levels are extremely low (Crosthwaite et al., 1997), and the ectopic expression of WC proteins in *wc* mutant backgrounds quickly leads to the induction of *frq* transcription (Cheng et al., 2001b), indicating that the WC proteins are activators of *frq* transcription. In constant darkness, D-WCC activates *frq* transcription by binding directly to the C-box (the distal LRE, containing two GATG repeats) in the *frq* promoter (Froehlich et al., 2003). *In vitro* DNA-binding assays have shown that D-WCC binds the C-box rhythmically and the disruption of the C-box leads to circadian arrhythmicity, indicating that the activation of *frq* transcription by D-WCC is essential for clock function (Froehlich et al., 2003). The binding of the D-WCC to the *frq* LREs was also confirmed *in vivo* using a chromatin-immunoprecipitation assay (He and Liu, 2005).

WCC, that is reconstituted *in vitro*, can bind to *frq* LREs in gel-shift assays, indicating that the WCC does not require additional components for

DNA binding (Froehlich et al., 2003; He et al., 2005b). The activation of frq transcription in the dark requires the Zn-finger DNA-binding domains of both WC proteins, since the deletion or mutation of their respective Zn-fingers abolishes the expression of frq in the dark (Cheng et al., 2002; Collett et al., 2002; Crosthwaite et al., 1997). Thus, the binding of the D-WCC to the C-box of frq needs the DNA-binding domains of both WC-1 and WC-2, with each domain perhaps binding to one of the GATG motifs. In addition to the LOV domain of WC-1, only the DNA-binding domain of WC-2 but not that of WC-1 is required for light-induced transcription (Cheng et al., 2003b). Thus, since the WC-1 LOV domain is dispensable for frq activation in the dark, the light and dark functions of WC-1 can be molecularly separated (He et al., 2002).

C. FRQ and FRH: The negative elements in the circadian negative feedback loop

Neurospora strains that carry mutations at the frq locus were some of the first circadian clock mutants identified in eukaryotes and it was soon established that frq plays a central role in the Neurospora circadian clock (Feldman and Hoyle, 1973; Loros et al., 1986). First, mutations in the frq gene result either in short-period, long-period (ranging from 16 to 35 h), or arrhythmic phenotypes, and can impair temperature compensation of the clock (Aronson et al., 1994b; Liu et al., 2000). Second, the deletion of frq abolishes circadian rhythmicity and temperature compensation of the circadian clock (Aronson et al., 1994b; Loros and Feldman, 1986). Third, constitutive expression of frq results in arrhythmicity (Aronson et al., 1994a), indicating that its rhythmic expression is essential for clock function. Fourth, FRQ is a state variable of the Neurospora clock and changes in frq mRNA and protein levels can phase-shift the clock, a process that mediates resetting and entrainment of the clock by light and temperature (Crosthwaite et al., 1995; Huang et al., 2006; Liu et al., 1998). Finally and most importantly, FRQ represses its own transcription (Aronson et al., 1994a; Merrow et al., 1997). In a frq^9 strain (a strain making a truncated, nonfunctional FRQ due to a frameshift mutation), frq levels are constantly high. On the other hand, the ectopic expression of frq represses expression of the endogenous frq gene. Together, these properties define the Neurospora circadian negative feedback loop in which FRQ is a central player as a negative element (Dunlap, 1999; Loros and Dunlap, 2001).

Full-length FRQ protein is 989 amino acids in length with a coiled-coil domain located near the N-terminal part (Cheng et al., 2001a). FRQ dimerizes through its coiled-coil domain, and FRQ dimerization is required for its interaction with the WCC and essential for clock function. In vivo, two alternatively translated forms of FRQ, large FRQ (lFRQ) and small FRQ (sFRQ), that differ by 99 amino acids at the N-terminus are made due to the use of two in frame AUGs (AUG1 and AUG3) (Garceau *et al.*, 1997; Liu *et al.*, 1997), the result of alternative mRNA-splicing events (Colot *et al.*, 2005; Diernfellner *et al.*, 2005). Small amounts of FRQ are found in the nucleus and its nuclear localization is essential for its circadian clock function (Luo *et al.*, 1998). However, the vast majority of FRQ is cytoplasmic and its function in the cytoplasm is unknown (Cheng *et al.*, 2005; Luo *et al.*, 1998; Schafmeier *et al.*, 2005). Both *frq* mRNA and FRQ protein show robust daily rhythms in constant darkness, peaking in the early morning and midday, respectively (Aronson *et al.*, 1994a; Garceau *et al.*, 1997).

FRH is the other core component of the Neurospora circadian negative feedback loop (Cheng *et al.*, 2005). FRH was discovered as an FRQ-interacting protein after purification of FRQ from Neurospora. It encodes a protein of 1106 amino acids with a DEAD/DEAH box and a helicase C domain, both of which are found in members of the DExD/H family of RNA helicases. FRH belongs to the SKI2 subfamily of RNA helicases that are found in organisms as diverse as fungi and mammals. Its homologue in *Saccharomyces cerevisiae*, Dob1p/Mtr4p, has been shown to be an essential nuclear cofactor of the yeast exosome complex, which is an important regulator of RNA metabolism in eukaryotes (Allmang *et al.*, 1999; de la Cruz *et al.*, 1998; Hilleren and Parker, 2003; Jacobs *et al.*, 1998; Mitchell and Tollervey, 2000; Torchet *et al.*, 2002).

Immunodepletion assays showed that all FRQ is in a complex with FRH *in vivo*, while only about 40% of FRH forms a complex with FRQ (Cheng *et al.*, 2005), indicating that the levels of FRQ determine the amount of FFC in Neurospora. Unlike other core clock components in circadian negative feedback loops, *frh* is an essential gene, probably due to its essential role in mediating exosome functions. An inducible RNA interference approach was developed to examine the role of *frh* in the circadian clock (Cheng *et al.*, 2005). Down-regulation of *frh* by inducibly expressing an *frh* hairpin RNA in Neurospora results in low levels of FRQ and overt and molecular arrhythmicity, indicating that it is an essential clock component. As a consequence of the reduction in FRQ levels, down-regulation of *frh* induces high levels of *frq* mRNA, indicating that FRH is essential for the function of the negative feedback loop. FRH also associates with WCC and such interaction is independent of FRQ. Furthermore, the disruption of the FRQ–FRH interaction also abolishes the FRQ–WCC association, suggesting that FRH mediates the FRQ–WCC interaction. Together, these results establish FRH as another essential component in the circadian oscillator that works together with FRQ to establish the negative limb of the circadian feedback loop.

RNA helicases are known to unwind RNA and regulate RNA metabolism. Most of the FFC is found in the cytoplasm with unknown functions.

Since the FRH homologue in yeast, Dob1p/Mtr4p, has been shown to bind RNA and function as an essential cofactor for the exosome, it is also likely that the FFC may associate with RNAs and regulate their functions. Although the role of FRH as an RNA helicase in the circadian regulation is unclear at the moment, it is possible that the rhythmic fluctuation of the FRQ–FRH complex may be a mechanism by which the Neurospora circadian clock directly controls RNA processing and degradation. Interestingly, NONO, an RNA-binding protein, was recently found to be associated with the mammalian PER1 and is important for clock functions in both mammalian cell culture and fly (Brown et al., 2005).

D. Inhibition of WCC by FFC and the role of WC phosphorylation

Comparable to the circadian negative elements in Drosophila and mammals, FFC interacts with the positive element, the WCC, to inhibit the activity of WCC and close the circadian negative feedback loop (Cheng et al., 2001a; Denault et al., 2001; Merrow et al., 2001). The FRQ–WCC interaction needs both WC-1 and WC-2 proteins (Cheng et al., 2003b; Denault et al., 2001), and the disruption of the FRQ–WCC interaction abolishes FRQ's role in negative feedback regulation (Cheng and Liu, unpublished results; Cheng et al., 2001a).

It is unclear how FFC inhibits WCC activity. One possibility is that the FFC–WCC interaction may physically remove the WCC from DNA. However, this is not supported by the stoichiometry of the levels of the FFC, the WCC, and the FFC–WCC and their cellular localizations. WC-1 and WC-2 reside mostly in the nucleus (Schwerdtfeger and Linden, 2000; Talora et al., 1999) and WC-1 is the limiting factor for the formation of the WCC (Cheng et al., 2001b; Denault et al., 2001). Although the total amount of FRQ is comparable that of WC-1 in the cell, the vast majority (>90%) of FRQ is in the cytoplasm (Cheng et al., 2001a, 2005; Luo et al., 1998). More importantly, biochemical purification of FRQ and WCC indicates that only a small amount of FRQ is in a complex with the WCC and only a small fraction of the WCC pool associates with FRQ in vivo (Cheng et al., 2005; He et al., 2002; Schafmeier et al., 2005). Such a small amount of nuclear FFC cannot effectively inhibit the WCC via a simple physical interaction but may do so by recruiting an enzyme-catalyzed reaction to inhibit WCC.

Similar to FRQ, both WC proteins are phosphorylated in vivo (He et al., 2005b; Schafmeier et al., 2005; Schwerdtfeger and Linden, 2000; Talora et al., 1999). WC proteins are phosphorylated in constant darkness and become hyperphosphorylated after light exposure. Using mass spectrometry and purification of the WCC from Neurospora, five major in vivo WC-1 phosphorylation sites, located immediately downstream of the WC-1 Zn-finger DNA-binding domain, were identified (He *et al.*, 2005b). Mutation of these phosphorylation sites revealed that they are light independent and are critical for circadian clock function in the dark: Neurospora strains that carry mutations in these sites show short-period, low-amplitude, or arrhythmic phenotypes. In addition, despite its low WC-1 levels, normal or slightly higher levels of *frq* mRNA and FRQ proteins, as well as an earlier activation of *frq* transcription were observed in these mutants. These data suggest that the phosphorylation of WC-1 at these sites negatively regulates its activity and that WC phosphorylation is a critical step in the circadian regulation.

The importance of WC phosphorylation for circadian clock function was confirmed by the surprising observation by Brunner and colleagues that the light-independent WC phosphorylation is dependent on FRQ (Schafmeier *et al.*, 2005). In a *frq* null strain, both WC-1 and WC-2 are hypophosphorylated and their normal phosphorylation is restored on ectopic FRQ expression. Importantly, the activation of *frq* transcription correlates with the hypophosphorylation of the WCs. The partially purified FFC complex is not able to directly inhibit WCC binding to the C-box *in vitro*, ruling out the possibility that FFC inhibits WCC by simple physical interaction. It was demonstrated that the dephosphorylation of the Neurospora WCC significantly promotes its binding to the LREs (He and Liu, 2005). Taken together, these results suggest a model in which FFC closes the circadian negative feedback loop by inhibiting WCC activity via the promotion of phosphorylation of WC proteins. In this model, FFC may function as a substrate recruiting subunit of a kinase(s), via its physical interaction with the WCC, to phosphorylate WC-1 and WC-2 and inhibit their DNA-binding activity.

To confirm this model, the identification of the kinase(s) involved and the *in vitro* reconstitution of the FFC-dependent WCC inhibition are required. CK-1a and CKII are the FFC-dependent kinases that mediate the phosphorylation of WC-1 and WC-2 to inhibit the WCC activity *in vitro* and *in vivo* (Cheng *et al.*, 2005; Gorl *et al.*, 2001; He *et al.*, 2006). Previously, protein kinase C was suggested to phosphorylate the C-terminus of WC-1, thereby down-regulating WC-1 levels and regulating its functions in light responses (Franchi *et al.*, 2005). However, it is not clear whether PKC can phosphorylate WC-1 *in vivo* in the dark and whether it plays a role in the circadian negative feedback loop. Finally, the proline residues next to two of the five identified WC-1 phosphorylation sites suggest that their phosphorylation may be mediated by a proline-directed kinase (He *et al.*, 2005b). Interestingly, glycogen synthase kinase-3 (GSK-3), which has been shown to regulate the circadian clock in Drosophila (Martinek *et al.*, 2001), is a known proline-directed kinase. It is entirely possible that several kinases, with one being the priming kinase and dependent on FFC, work together to phosphorylate the WCs and regulate their activity.

E. FRQ phosphorylation regulated by kinases and phosphatases determines its stability and is important for its role in the circadian negative feedback loop

As soon as FRQ proteins are synthesized, they become phosphorylated and are progressively phosphorylated over time (Garceau et al., 1997). Thus, the amount and phosphorylation pattern of FRQ proteins show robust circadian rhythms in constant darkness. FRQ is mostly hypophosphorylated in the early subjective morning and becomes hyperphosphorylated in the subjective night before its levels decrease. An early study of FRQ phosphorylation suggested that phosphorylation promotes FRQ degradation and is an important determinant of period length of the clock (Liu et al., 2000). Subsequent studies have revealed that FRQ phosphorylation plays several important roles in the regulation of its circadian function, and the kinases and phosphatases which control FRQ phosphorylation status are important components of the Neurospora circadian oscillator (Gorl et al., 2001; He et al., 2003; Liu, 2005; Liu et al., 2000; Yang et al., 2001, 2002, 2003, 2004). Currently, three FRQ kinases (CKI, CKII, and CAMK-1) and two phosphatases (PP1 and PP2A) have been identified to regulate FRQ phosphorylation. Among the kinases, CKII appears to be the most important one, being an essential clock component, while PP1 and PP2A seem to have distinct roles within the clockwork.

1. Casein kinase 1a

The Drosophila DOUBLETIME (DBT) and the mammalian CKI homologues, CKI epsilon and CKI delta, phosphorylate PER proteins and play an important role in regulating period length of the insect and mammalian circadian oscillator (Kloss et al., 1998; Lowrey et al., 2000; Price et al., 1998; Xu et al., 2005). The conservation of the role of CKI in eukaryotic systems prompted Gorl et al. (2001) to examine its role in the Neurospora circadian clock. Among CK-1a and CK-1b, the two CKI homologues that exist in Neurospora, it is CK-1a that is more similar to DBT. Both CK-1a and CK-1b can phosphorylate the PEST domains of FRQ in vitro, but they cannot phosphorylate a region containing three previously identified phosphorylation sites. Importantly, CK-1a was found to associate with FRQ, suggesting that FRQ is also a CK-1a substrate in vivo. The CK-1a–FRQ association was further confirmed by the purification of FRQ (Cheng et al., 2005). Deletion of the PEST-1 domain leads to the reduction of FRQ phosphorylation species and FRQ degradation rate, resulting in a long-period rhythm. These data suggest that the phosphorylation of this region by CK-1a may regulate FRQ stability. We now know that CK-1a not only phosphorylates FRQ in vivo, but also mediates the FFC-dependent phosphorylation of WC-1 and WC-2 to inhibit the WCC activity (He et al., 2006). On the other hand, disruption of the ck-1b gene does not result in significant changes in the FRQ phosphorylation profile and FRQ oscillation despite its important role in growth and development, indicating that CK-1b is not essential for a functional Neurospora circadian clock (Yang *et al.*, 2003).

2. Calmodulin kinase 1

CAMK-1 is the first potential FRQ kinase that was identified by a biochemical purification approach (Yang *et al.*, 2001). *In vitro*, Ca/CaM-dependent kinase activity accounts for most of the kinase activities that phosphorylate FRQ. CAMK-1 strongly phosphorylates the FRQ region including amino acids 501–519. Disruption of *camk-1* in Neurospora leads to transient growth and developmental defects, but the quick reversal of *camk-1* mutants to the wild-type phenotype suggests that other kinases may replace the function of CAMK-1. The reversion of *camk-1* mutants prevents a firm conclusion about the *in vivo* role of CAMK-1 in phosphorylating FRQ, but the phase, period length, and light-induced phase shifting of the circadian clock were all modestly affected in the *camk-1* reversion mutants.

3. Casein kinase II

CKII is an FRQ kinase biochemically purified from the *camk-1* mutant strain (Yang *et al.*, 2002). In Neurospora, the disruption of *cka*, the only catalytic subunit of CKII, is not lethal (unlike in other eukaryotic organisms) but leads to severe growth and developmental defects. In the *cka* mutants, the FRQ phosphorylation pattern is significantly altered and FRQ proteins are mostly hypophosphorylated, indicating that FRQ is an *in vivo* substrate of CKII. In addition, the levels of FRQ are significantly higher in the *cka* mutant compared to a wild-type strain, suggesting that, in the wild type, FRQ phosphorylation by CKII promotes its degradation (Yang *et al.*, 2002).

　　In the *cka* mutant, the levels of FRQ protein are constitutively high and hypophosphorylated and circadian rhythms of *frq* and several *clock-controlled genes* (*ccgs*) are abolished in constant darkness, indicating that CKII is an essential component in the Neurospora circadian system. In contrast to the prediction that high levels of FRQ should lead to the repression of *frq* RNA, the levels of *frq* RNA in the *cka* mutant randomly fluctuate at high levels, suggesting that the circadian negative feedback loop is impaired in the mutant and CKII phosphorylation of FRQ is required for its role as a transcriptional repressor. Furthermore, WC complex was found to preferentially interact with the hypophosphorylated FRQ in the *cka* mutant. Thus, phosphorylation of FRQ by CKII appears to have at least three roles in the clock: promoting the degradation of FRQ, inhibiting the FRQ–WC interaction, and mediating the completion of the circadian negative feedback loop (Yang *et al.*, 2002). The latter was recently shown due to the role of CKII in mediating the FFC-dependent phosphorylation of WC-1 and WC-2, which inhibits WCC activity (He *et al.*, 2006).

The important roles of CKII in the Neurospora clock were further supported by the disruption of one of the two CKII regulatory subunit genes, *ckb1* (Yang et al., 2003). In the *ckb1* mutant, FRQ proteins are hypophosphorylated and more stable than in the wild-type strain, and circadian rhythms of conidiation and FRQ protein oscillation were observed to have long periods but low amplitudes. In addition, mutations of several FRQ CKII phosphorylation consensus sites result in hypophosphorylated FRQ and long-period rhythms. Interestingly, CKII was also later shown to be an important clock component in Drosophila where it phosphorylates PER and thereby potentiates the repressor function of PER in the circadian negative feedback loop (Nawathean and Rosbash, 2004).

4. Protein phosphatases 1 and 2A (PP1 and PP2A)

PP1 and PP2A are two major classes of protein phosphatases in eukaryotic organisms. To carry out their diverse cellular functions, protein phosphatases have a large number of regulatory proteins. Both PP1 and PP2A are important regulatory components in the Neurospora clock (Yang et al., 2004). In a partially functional *ppp-1* (encoding the catalytic subunit of PP1) mutant strain, which shows significantly reduced PP1 activity due to a missense mutation, FRQ is less stable, and this instability results in a short-period phenotype and a significant phase advance of circadian conidiation. On the other hand, disruption of the Neurospora *rgb-1* gene (encoding for a regulatory subunit for PP2A) does not affect the stability of FRQ. Nonetheless, in the *rgb-1* mutant, the levels of FRQ protein and *frq* mRNA are low, and the clock oscillates with a low-amplitude and long-period rhythm. In addition, PP1 and PP2A can dephosphorylate FRQ *in vitro*, and the FRQ phosphorylation profile is altered in the *rgb-1* mutant strain *in vivo*. Together, these data suggest that PP1 and PP2A play distinct roles in the Neurospora circadian clock and counterbalance the effects of the kinases. By dephosphorylating FRQ, PP1 regulates the stability of FRQ, whereas PP2A has important functions within the negative feedback loop. RGB-1 was also shown to regulate the phosphorylation of WC-1, a process which may affect the activity of the WCC (Schafmeier et al., 2005). TWS, the homologue of RGB-1, was also shown to be an important clock component of the fly circadian oscillator (Sathyanarayanan et al., 2004).

F. Degradation of FRQ through the ubiquitin-proteasome system requires a conserved SCF-type ubiquitin ligase, SCF^{FWD-1}

The phosphorylation-dependent degradation of FRQ is similar to that of PER proteins in Drosophila and mammals and is mediated by the ubiquitin/proteasome (Eide et al., 2005; Grima et al., 2002; He et al., 2003; Ko et al., 2002). FRQ is ubiquitinated *in vivo*, and its proper degradation requires FWD-1, an F-box/ WD-40 repeat-containing protein and the Neurospora homologue of the Drosophila protein Slimb (He *et al.*, 2003). F-box proteins function as the substrate-recruiting subunit of the SCF-type ubiquitin (E3) ligases. The disruption of *fwd-1* results in the accumulation of hyperphosphorylated FRQ, indicating that FRQ phosphorylation precedes its degradation. In the *fwd-1* mutant, the circadian rhythms of FRQ and other clock-controlled gene expression and the circadian conidiation rhythms are abolished. Furthermore, FWD-1 without its F-box forms a stable complex with FRQ *in vivo*, strongly suggesting that FWD-1 is the substrate-recruiting subunit of the SCF^{FWD-1} complex that mediates FRQ ubiquitination and degradation. The absence of a stable interaction between FRQ and FWD-1 in a wild-type strain suggests that the FWD-1–FRQ interaction is transient and FRQ is rapidly ubiquitinated and degraded by the proteasome. Taken together, these data indicate that the progressive phosphorylation of FRQ at multiple independent sites fine-tunes the stability of FRQ and determines the period of the clock. In this model, extensive FRQ phosphorylation increases its affinity with the WD-40 domain of FWD-1 and leads to FRQ degradation. Thus, the progressive phosphorylation events in fungi, flies, and mammals may provide a highly regulated time delay in the circadian negative feedback loop to achieve stable 24-h rhythmicity.

The COP9 signalosome (CSN), a conserved multisubunit complex in all eukaryotes, was shown to be important for clock function by regulating the stability of the SCF^{FWD-1} complex in Neurospora (He *et al.*, 2005a). The disruption of CSN in Neurospora dramatically destabilizes the SCF^{FWD-1} complex, a result of its enhanced autoubiquitination due to the lack of deneddylation of CULLIN-1 (a component of SCF complexes) by CSN. This leads to low levels of SCF^{FWD-1} complex and impaired FRQ degradation and clock functions. Interestingly, the loss of CSN function reveals an FRQ-independent oscillator which drives a long-period and temperature-compensated conidiation rhythm that persists in DD, LL, and temperature cycles.

G. Conservation of eukaryotic circadian oscillators, from Neurospora to animals

Remarkable conservation, both at the mechanistic level and at the level of clock proteins, exists between the circadian oscillator of Neurospora and those of insects and mammals.

First, a similar architecture of the circadian oscillator is shared from Neurospora to mammals. The Neurospora, Drosophila, and mammalian circadian oscillators are all based on autoregulatory circadian negative feedback loops (Dunlap, 1999; Young and Kay, 2001). In all of these self-regulatory circuits, heterodimeric complexes consisting of two PAS domain-containing transcription factors function as the positive elements that activate the negative components of the loop. The negative factors, in turn, close the feedback cycle by physically interacting with the positive elements. Furthermore, the interlocked positive feedback loops that are found in these circadian systems appear to play a similar role in clock functions, that is provide robustness to the feedback cycle (Cheng et al., 2001b; Glossop et al., 1999; Kim et al., 2002; Lee et al., 2000; Preitner et al., 2002; Shearman et al., 2000). WC-1 shows some homology to BMAL1, a bHLH transcription factor that plays a role similar to WC-1 in the mammalian circadian feedback loop (Lee et al., 2000; Tauber et al., 2004). Moreover, and perhaps similar to the association of FRH with FRQ in Neurospora, an RNA-binding protein was found to be associated with PER proteins to regulate clock function in mammals and Drosophila (Brown et al., 2005). Thus, although the core components of the Neurospora circadian oscillator are mostly not sequence homologues of those in insects and mammals, functionally, they all have very similar roles to their animal counter parts in the generation of circadian rhythmicity.

     Conservation is also seen in the regulation of clock proteins. Like FRQ and WCs, the core components of the animal clock are regulated by phosphorylation. Similar to FRQ, PER proteins are progressively phosphorylated and phosphorylation promotes their proteasome-mediated degradation and is important for the repressor function of PER (Young and Kay, 2001). FRQ and the PER proteins, although not sequence homologues, are both phosphorylated by CKI and CKII (Akten et al., 2003; Gorl et al., 2001; Kloss et al., 1998; Lin et al., 2002; Lowrey et al., 2000; Nawathean and Rosbash, 2004; Price et al., 1998; Xu et al., 2005; Yang et al., 2002, 2003), and can be dephosphorylated by the same phosphatases (Sathyanarayanan et al., 2004; Yang et al., 2004). Furthermore, a conserved SCF E3 ligase mediates the phosphorylation-dependent ubiquitination and degradation of both FRQ and PER (Eide et al., 2005; Grima et al., 2002; He et al., 2003; Ko et al., 2002). The conservation of these posttranslational clock regulators suggests that they may be the common evolutionary link among eukaryotic circadian oscillators. On the basis of this observation, it was proposed that CSN, a highly conserved eukaryotic regulator in protein degradation, may be another shared gear in the circadian clock of higher eukaryotes and its dysfunction may contribute to the circadian and sleep phenotypes of human Smith Magenis Syndrome (He et al., 2005a).

IV. TEMPORAL INPUT FROM THE ENVIRONMENT

The core circadian clock components are part of a biological clockwork that can produce molecular and physiological rhythms in the absence of external time cues. Under these free-running conditions, the clock is self-excitatory, that is feeds back on itself to create a rhythmic environment within the cell. To stay in synchrony with the natural world, circadian clocks must be coupled to the rhythm of the environment such that a stable phase relationship between clock and the environment is established. This coupling process is called entrainment (Johnson *et al.*, 2003). Only in the entrained circadian system can cellular functions happen at their optimal time everyday and throughout the year. Naturally, the daily rhythms in light and temperature conditions have become the most important signals from which the molecular clockwork takes its time cues. A simplified summary of input signals that feed into the circadian oscillator is given in Fig. 2.2.

A. Light input into the circadian clock

1. Early light signaling events

Within the visible spectrum of light, blue light is generally the most important for the entrainment of circadian clocks (Gehring and Rosbash, 2003). Not only is Neurospora's clock blue-light sensitive but in fact, all of Neurospora's described light responses are blue-light dependent (Linden, 2002; Liu *et al.*, 2003). Blue-light perception in Neurospora requires a heteromeric WCC, consisting of WC-1 and WC-2, which we have already introduced as a key component of the FFC-WCC oscillator (Ballario *et al.*, 1996; Crosthwaite *et al.*, 1997; Linden and Macino, 1997; Talora *et al.*, 1999). Moreover, WC-1 protein has been shown to be the main blue-light receptor in Neurospora, further underlining the tight link between light input and the circadian oscillator in Neurospora (Froehlich *et al.*, 2002; He *et al.*, 2002). This observation is paralleled by similarly close connections between sensory input and core oscillators in Drosophila (Krishnan *et al.*, 2001).

WC-1 gains photoreceptor status via its LOV domain, a specialized PAS domain that harbors flavin as a chromophore (Froehlich *et al.*, 2002; He *et al.*, 2002). The mechanism of chromophore attachment is similar to other LOV-type PAS domains that have been implicated in blue-light reception and which contain a signature motif for flavin binding (Cheng *et al.*, 2003a; Crosson and Moffat, 2001). Within the LOV domain, the flavin (FAD or FMN) forms a blue-light-dependent adduct with a conserved cysteine residue in the LOV domains (Crosson and Moffat, 2002; Crosson *et al.*, 2003). We lack biophysical information on the early events that follow the absorption of a blue-light photon by WC-1, but examples for LOV domain-mediated blue-light signaling from other eukaryotic model systems give clues as to the possible mechanism. For example, we know from structural studies of the plant PHY3 LOV2 domain that blue-light absorption drives the formation of a reversible covalent bond between a conserved Cys residue in LOV2 and the flavin cofactor. This adduct formation

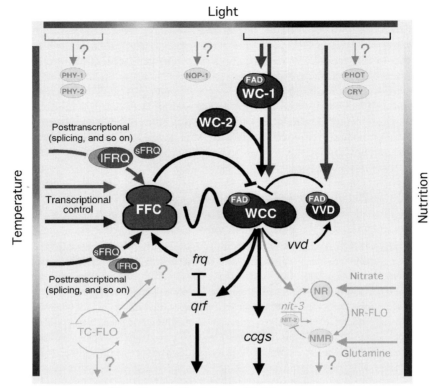

Light

Circadian rhythmicity and light responses

Figure 2.2. Schematic representation of the Neurospora circadian system with emphasis on the environmental input (light, temperature, nutrition) pathways that feed into the core oscillator. Known pathways are depicted in bold and color, whereas components with unknown or controversial functions are depicted in light gray. A simplified version of the central feedback loop (as depicted in more detail in Fig. 2.1) is shown at the center of the system. Light perception in Neurospora is probably mediated by a number of photoreceptors that absorb wavelength across the entire spectrum of visible light, but only the WC-1 and VVD proteins are established as blue-light photoreceptors, which seems to mediate all known light responses in Neurospora. WC-1 is the primary photoreceptor and is required for the expression of VVD, which the serves as a secondary photoreceptor and repressor of WCC activity. The VVD feedback loop mutes light responses in Neurospora and impacts on light resetting and entrainment of the Neurospora circadian clock (for more information see text). Similarly, the expression of *qrf*, an antisense transcript to the *frq* mRNA, mutes light resetting of the Neurospora clock via an unknown mechanism. Temperature is sensed via transcriptional and posttranscriptional responses that act on the *frq* gene and its products. At higher temperatures, more large FRQ (lFRQ) is made, whereas at lower temperatures, relatively more small FRQ (sFRQ) is present. The overall quantity of FRQ and the

causes the planar flavin to tilt slightly and is believed to represent the active signaling state of the molecule. The photoinduced tilt of the flavin chromophore seems to induce slight, but significant motions in the backbone of the polypeptide chain of PHY3 LOV2 that "push" on a conserved surface salt bridge. The photoinduced conformational change is predicted to lead to a corresponding change in interaction probability with downstream signaling components, thus triggering signal transduction (Crosson and Moffat, 2002; Crosson *et al.*, 2003; Harper *et al.*, 2003). The structure of the LOV domain of WC-1 is not currently available, and thus prediction of the early signaling event is speculative. However, significant similarities to the solved structures of LOV domains are likely, given their similar function, sequence identity, and type of cofactors. Compared to the LOV domain of PHY3, WC-1 contains additional amino acids which are predicted to lie close to or within the structure that is proposed to be involved in blue-light-mediated structural changes of PHY3 LOV2. This additional loop in the WC-1 amino acid sequence might accommodate the FAD that is bound to WC-1 or may relay blue-light-mediated structural changes to downstream targets and perhaps relates to the very long photocycle of WC-1 (He and Liu, 2005; He *et al.*, 2002; Heintzen, 2005). Both features, that is extra loop and long photocycle, have also been reported for FKF1, a flavin-binding LOV domain protein that imposes cyclic degradation of a transcriptional repressor of CO, a key factor of photoperiodic flowering control in *Arabidopsis* (Cheng *et al.*, 2003a; Imaizumi *et al.*, 2003, 2005).

It is also conceivable that light-induced conformational changes might reveal previously hidden phosphorylation sites or may render WC-1 a better substrate for kinases or phosphatases (Heintzen, 2005). This would be consistent with the observation that WC-1 and WC-2 undergo light-induced changes in phosphorylation status (Schwerdtfeger and Linden, 2000; Talora *et al.*, 1999).

---

ratios of lFRQ to sFRQ are largely regulated at the posttranscriptional level via temperature-sensitive alternative splicing mechanisms and temperature-dependent ribosome scanning mechanisms (see text for more detail). The environmental signals are integrated at the level of the oscillator and control a large number of clock-controlled genes (*ccgs*) that impart rhythmicity on the physiology of the whole organism. Little is known about nutritional input into the circadian oscillator. A few FRQ-less oscillators that become visible or operate in the absence of the circadian FRQ-WCC oscillator have been described of which only two are depicted here. The nitrate-reductase FLO (NR-FLO) is predicted to center around the pathways of nitrate assimilation and only present when nitrate is the sole nitrogen source but is suppressed by ammonium or glutamine. The components of a temperature-controlled FLO (TC-FLO) that regulates rhythmic conidiation in temperature cycles are unknown. Whether these FLOs are controlled by circadian oscillators and contribute to overt circadian rhythmicity in Neurospora has not been established or is controversial. (See Color Insert.)

Thus, phosphorylation and dephosphorylation are likely to be key events in activating or repressing the activity of the WCC. WC-1, in particular, becomes strongly hyperphosphorylated on illumination, but quickly returns to the hypophosphorylated forms in extended periods of illumination (He et al., 2005a; Heintzen et al., 2001; Schwerdtfeger and Linden, 2001; Talora et al., 1999). The transient nature of WC-1 hyperphosphorylation correlates well with the transient nature of Neurospora light responses, a phenomenon known as photoadaptation. On the basis of this correlation, it was proposed that it is hyperphosphorylated WC-1 which activates transcription of target genes (Talora et al., 1999). However, more recent work suggests that hyperphosphorylated WC-1 is transcriptionally inactive and the hypophosphorylated WC-1 is in fact the signaling active form (He and Liu, 2005). This evidence is based in part on high resolution RNA and WC-1 kinetics that demonstrate that transcript levels of light-induced genes rise before hyperphosphorylation of WC-1 becomes evident and have already started to fall by the time WC-1 becomes maximally phosphorylated. More importantly it was shown that hyperphosphorylated WC-1 binds less strongly to target promoters and dephosphorylation of WC-1 increases promoter binding. This situation is very similar to that observed for the WCC functioning in the dark, where it is also the hypophosphorylated WCC that is transcriptionally active (He et al., 2005b; Schafmeier et al., 2005).

The early light signaling events may be envisioned as follows: on blue-light illumination, the WCC undergoes a rapid conformational change, transforming the WCC into a large photoactivated complex (L-WCC) that induces transcription. The conformational change may also promote L-WCC phosphorylation, thus inactivating the complex and targeting its constituents for degradation. Previous work has shown that WC-1 can only exist in a WCC (Cheng et al., 2003b). Perhaps light-induced hyperphosphorylation destabilizes the WCC, deactivating it and exposing additional phosphorylation sites that further promote the degradation of both transcription factors. Indeed, several studies have indicated that WC-1 phosphorylation triggers its degradation (He and Liu, 2005; Lee et al., 2000; Talora et al., 1999). In addition, dominant negative mutants of PKC, a likely WC-1 kinase in vivo, accumulate higher levels of WC-1 whereas the constitutive overexpression of PKC reduces levels of WC-1 (Franchi et al., 2005).

How are these light signaling events linked to the clock? One of the targets of the photoactivated WCC is the frq gene (Crosthwaite et al., 1997). Two light responsive elements (LREs) have been identified in the frq promoter that are bound by the WCC in vivo (Froehlich et al., 2002, 2003; He and Liu, 2005). The distal element is also bound by the WCC in the dark and helps to maintain rhythmic frq expression in constant conditions (Froehlich et al., 2003). However, the L-WCC that assembles on the frq promoter in the light is larger than the D-WCC. In vitro studies have shown that WC-1 and WC-2 are the only required components for light-dependent binding to target promoters, suggesting that the large light-complex is not the result of, or does not require, recruitment of other factors (Froehlich *et al.*, 2002; He and Liu, 2005). Instead, it seems that WC-1 homodimerizes *in vivo* with the potential to form a large trimeric WC-1:WC-1:WC-2 complex (Cheng *et al.*, 2003b). It is possible that the L-WCC binds to the *frq* promoter in the light, whereas the dimeric D-WCC complex functions in the dark. It is not clear whether the L-WCC exists in the dark or only forms from the D-WCC after light exposure. Whatever the mechanism, the link between light sensing and transcriptional activation of the *frq* gene provides a simple explanation of how light signals can influence the clockwork.

2. Light resetting and entrainment

Crosthwaite *et al.* (1995) proposed a simple model based on molecular evidence to explain light resetting in eukaryotes. In this model, rapid light-induction of *frq* RNA explains the opposing effects of light on the clock. On the basis of the circadian rhythm of *frq* transcript levels in the dark, the model proposes that light will act on the rising slope of the *frq* cycle (i.e., during subjective dawn) by allowing *frq* levels to peak earlier and thus advancing the clock cycle, while light acting on the falling slope (at subjective dusk) will slow down the fall of *frq* levels and thus delay the cycle. In contrast, during the subjective day, when *frq* levels are peaking, a further rise in *frq* levels will elicit small or no effect on the cycle kinetics, leaving the clock unperturbed (Crosthwaite *et al.*, 1995).

Similar principles have been uncovered in other circadian systems, where the rapid light-triggered induction or degradation of central clock components can be used to explain resetting behavior in these organisms (Ashmore and Sehgal, 2003; Shigeyoshi *et al.*, 1997; Young and Kay, 2001). Interestingly, *qrf*, an antisense transcript arising from the *frq*-locus, whose levels cycle in antiphase to *frq* transcript levels is also strongly light-induced. Strains in which light-induction of *qrf* is absent show a stronger resetting response to light pulses but have a normal circadian clock in the dark, suggesting that *qrf* (and similar to VVD, see below) mutes the light response of the Neurospora circadian clock but does not directly impinge on the central clockwork (Crosthwaite, 2004; Kramer *et al.*, 2003).

Rapid changes in levels of clock components can explain light resetting by brief light pulses but they leave open the question of how full photoperiods, that is the more continuous action of light, facilitates circadian clock entrainment (Johnson *et al.*, 2004). It is possible that despite the continuous presence of light, only the dawn and dusk transitions are sensed (Johnson *et al.*, 2004). Blocking out the action of light during the day could be achieved via behavioral or molecular means including mechanisms by which the clock itself influences its light sensitivity, a process known as circadian gating (Dragovic *et al.*, 2002; Fleissner and Fleissner, 1992; Gillette and Mitchell, 2002; Heintzen *et al.*, 2001; McWatters *et al.*, 2000; Millar and Kay, 1996). In discrete models of entrainment, light acts to align the circadian clock to the real world by deleting or inserting portions of a given circadian cycle. Alternatively, the action of light may stretch or compress the circadian cycle, that is adjust the duration of the cycle in a more continuous fashion. Because period length of circadian clocks is often dependent on light intensity (Aschoff's rule), this rule is often quoted in support for models of continuous entrainment (Johnson *et al.*, 2004).

Recently, some progress has been made in testing models of entrainment in more detail in Neurospora. In particular, the study of *vivid* (*vvd*) mutants has shed some light on the molecular pathways that are important for entrainment to full photoperiods (Elvin *et al.*, 2005; Heintzen *et al.*, 2001). The *vvd* gene encodes a small PAS/LOV protein, most similar to the PAS/LOV domain of WC-1 (Heintzen *et al.*, 2001). VVD acts as a repressor of light responses and is a WC-dependent blue-light photoreceptor, that is, required for measuring changes in light intensities (Cheng *et al.*, 2003a; Heintzen *et al.*, 2001; Schwerdtfeger and Linden, 2001, 2003; Shrode *et al.*, 2001). The observation that VVD mutes WCC activity suggested a role in light resetting and entrainment of the Neurospora circadian clock. Moreover, because *vvd* transcript levels were shown to be transiently clock-controlled, the expression profile and action of VVD fulfilled the formal requirements for a molecular gating loop in which the clock would control its own light sensitivity by rhythmically regulating VVD levels. In line with these predictions, *vvd* knockout mutants show a much stronger resetting response to light pulses when compared to clock wild-type strains and display a significantly muted, albeit not abolished gating response (Heintzen *et al.*, 2001). The *vvd* mutants also display a roughly 4-h delay in the phase of conidiation in constant conditions (Elvin *et al.*, 2005; Heintzen *et al.*, 2001). Investigations into the reason for this phase phenotype have uncovered that VVD imposes phase control by muting light resetting at dawn, through its repressive action on WCC activity and promoting clock resetting at dusk, via accelerating the drop in *frq* RNA levels. Because, in the wild type, VVD mutes dawn resetting of the clock, the circadian clock can run almost unperturbed into the light. At the end of the day, the dusk phase of this day-oscillator, as determined by the FRQ levels, will then control the speed of feedback repression of *frq* and thus the phase of clock-controlled conidiation at night (Elvin *et al.*, 2005). In *vvd* mutants, this daytime oscillator collapses and with it any photoperiod-dependent phase information. In consequence, Neurospora must resume its oscillation with a fixed and significantly delayed phase. Given VVD's importance for light responses, it is easy to imagine how VVD impacts on the entrained circadian system in Neurospora when a range of photoperiods (from short days to long days) are used for entrainment. Under these conditions, VVD enables wild-type strains to adjust their phase of conidiation in a photoperiod-dependent manner such that conidiation occurs during the night or early morning. This ability is lost in *vvd* mutants, leading to extensive conidiation during the day in long photoperiods. In particular, VVD's action may help to protect spore development from the damaging effects of light and desiccation by ensuring clock function throughout the full light–dark cycle and thus maintaining a biologically advantageous clock phase.

3. Photoperiodism

It was Erwin Bünning who first suggested that circadian clocks could be used for measuring changes in daylength and serve as a basis for photoperiodism (Bünning, 1936). His external coincidence model proposes that the circadian clock imposes a rhythm in light sensitivity whose photosensitive phase must coincide with the external light stimulus to elicit a photoperiodic response. This model is now generally accepted as the basis for many (but not all) photoperiodic phenomena (Saunders, 2005). Evidence for photoperiodism in Neurospora was discovered when the production of sexual and asexual spores as well as levels of carotenoids were investigated in different photoperiods (Roenneberg and Merrow, 2001; Tan *et al.*, 2004b). By systematically varying the daylength, Tan *et al.* observed that the production of asexual conidiospores peaked around equinox, followed by peaks in the production of protoperithecia and carotenoids when days became longer. All of these photoperiodic phenomena are absent in *frq* null mutants suggesting that a functional circadian clock is required for the observed photoperiodic changes in sporulation and carotenogenesis.

At the molecular level, photoperiodic responses are best understood in plants where the products of the *constans* (*co*) gene have been identified as key components of the photoperiodic flowering response (Bünning, 1936; Suarez-Lopez *et al.*, 2001; Yanovsky and Kay, 2002) and have provided the first molecular support for the external coincidence model of photoperiodic responses. In Arabidopsis, the circadian clock phases the expression of the *co* gene such that in short days it is only expressed in the dark but, with increasing daylength, its expression and the presence of light eventually coincide. Thus, under long-day conditions, light-activated cryptochrome and phytochrome A stabilize CO, which in turn enables CO to activate genes required for flower development (Valverde *et al.*, 2004).

It seems possible that similar processes may contribute to photoperiodism in Neurospora. For example, when investigating *frq* RNA and FRQ protein levels in different photoperiods, Tan *et al.* (2004a) found that FRQ protein is not phased to the light or dusk transitions. This independency of FRQ oscillations on dusk and dawn transitions is in line with other observations that conclude that the Neurospora clock is functional across full light–dark cycles (Elvin *et al.*, 2005; Gooch, 1985; Tan *et al.*, 2004a) and not just passively driven by the light–dark cycle, a prerequisite for its use in photoperiodic control. Perhaps the entrained circadian clock in Neurospora facilitates photoperiodic responses by creating rhythms in photosensitivity as described for plants.

The results described above suggest that entrainment to full photoperiods may not be as simple as predicted from light-pulse experiments. Moreover, the use of skeleton photoperiods has revealed that brief light pulses presented at dawn and dusk cannot mimic the characteristic phase angle of entrainment reached in complete photoperiods in wild-type Neurospora (Elvin et al., 2005). In contrast, vvd mutants establish a similar phase angle in both skeleton and full photoperiods, suggesting that a more continuous action of light is required for entrainment to the natural environment, and that VVD mediates this continuous effect (Elvin et al., 2005).

The molecular mechanism by which VVD exerts its functions is unknown but is likely to involve direct or indirect effects of VVD on WCC phosphorylation/activity. For example, WC-1 is constitutively hyperphosphorylated in vvd mutants, perhaps indicating that VVD activates a phosphatase or represses a kinase (Heintzen et al., 2001; Schwerdtfeger and Linden, 2001). Hyperphosphorylation coincides with rapid WC-1 degradation and newly synthesized WC-1 is not detectable in vvd mutants (Heintzen et al., 2001; Schwerdtfeger and Linden, 2001). On the basis of its similarity to the WC-1 PAS/LOV domain, VVD may compete for WC-1 binding to WC-2 and thus repress light signaling. However, a number of facts speak against (but do not rule out) such a scenario: first, it is not the WC-1 PAS/LOV domain that mediates WC-1–WC-2 interaction (Ballario et al., 1998; Cheng et al., 2003b; He et al., 2002) and second, VVD is predominantly localized to the cytoplasm, whereas the WCC is nuclear (Schwerdtfeger and Linden, 2003).

On the basis of the observation that hyperphosphorylated WC-1 is the inactive form of WC-1, one would expect a decreased rather than an increased state of light signaling in vvd mutants. Therefore, the regulation of light signaling must be more complex possibly involving other factors or activities required for initial activation or, perhaps some transitional state(s) of the WCC conveys light signaling in Neurospora (Heintzen, 2005). More work is required to understand the molecular detail of entrainment but it seems likely that both discrete and continuous light responses establish the characteristic phase angle of entrainment in natural conditions (Johnson et al., 2004).

4. Other light input

Mining of the Neurospora genome sequence (www.broad.mit.edu/annotation/fungi/Neurospora_crassa_7) has revealed a number of potential light signaling proteins with similarities to phytochromes (PHY-1, PHY-2), cryptochrome (CRY), opsin (NOP-1), or LOV domain-containing proteins (PHOT)

(Froehlich *et al.*, 2005). This suggests that other blue and possibly red-light signaling pathways exist in Neurospora. Various laboratories have created gene knockouts for these components and analyzed single or multiple knockouts in respect to their effect on light responses and the circadian clock. So far none of these genes has been shown to have a major impact on the Neurospora circadian system or light responses in general (Froehlich *et al.*, 2005; Mark Elvin, Christian Heintzen, Ping Cheng, Yi Liu, unpublished results). However, CRY expression is light-regulated and may have roles that were previously assigned to the WCC (Froehlich *et al.*, 2005). None of the other putative photoreceptors shows light-dependent regulation. *phy-1* transcript abundance is regulated by the circadian clock; however, the levels of PHY-1, its phosphorylation status, and subcellular localization seem to be constitutive. In summary, any roles of these putative photoreceptors in light responses or circadian clock function remain elusive. Their role in light sensing may be subtle or involve photoresponses that are currently unknown in Neurospora.

B. Temperature input

Temperature affects the circadian system in three major ways: the most obvious effect is its action as an entrainment cue. In a 24-h temperature entrainment cycle and constant darkness, the relatively higher temperature is generally perceived as subjective day and the lower temperature as subjective night. Second, temperature imposes the limits under which the clock is operational, that is sets temperature-dependent thresholds above or below which the clock will not operate. The third influence of temperature on the circadian system is temperature compensation of the period of the clock. All three temperature effects on the circadian system have been studied in Neurospora.

1. Temperature sensing

The expression of *frq* is temperature regulated and could serve as the temperature-sensitive event, regulating the Neurospora circadian clock. After a temperature step up, *frq* mRNA level transiently increases, while a temperature step down results in its transient decrease, indicating a temperature-dependent transcriptional response (Liu, 2003; Liu *et al.*, 1998). Under constant temperature conditions, FRQ protein levels are higher at higher temperatures, whereas *frq* transcript levels stay more or less constant suggesting a posttranscriptional or posttranslational mechanism is largely responsible for temperature-sensitive FRQ expression (Colot *et al.*, 2005; Diernfellner *et al.*, 2005; Liu *et al.*, 1997, 1998). Earlier work has shown that two FRQ forms (lFRQ and sFRQ) arise from the alternative use of two in frame ATGs and the relative ratio of the two forms changes as temperature changes: high temperature preferentially promotes the expression of lFRQ resulting in a relative increase of sFRQ at lower temperatures (Liu *et al.*, 1997). Some evidence suggests that the choice of two alternative transcription start sites in the *frq* promoter is also temperature dependent, but this does not appear to influence the overall levels of *frq* transcripts in the cell (Colot *et al.*, 2005). The cause for the production of the two FRQ forms was recently shown to be a temperature-sensitive alternative splicing event at the *frq* 5′ UTR. Splicing events of the *frq* transcript are surprisingly complex and the major temperature-sensitive splicing event appears to involve a single intron that removes the first ATG of FRQ. Similar to temperature-regulated splicing mechanisms in other systems, the nonconsensus splice site renders splicing temperature sensitive. At low temperatures, the nonconsensus intron is preferentially spliced and removes the first ATG to produce only sFRQ. At higher temperatures, splicing is inefficient yielding relatively more lFRQ which arises from the unspliced *frq* transcript. The general increase in FRQ levels is, however, not controlled by temperature-sensitive splicing but instead can be partly explained by a temperature-dependent translational control mechanism mediated by uORFs in the *frq* 5′ UTR (Diernfellner *et al.*, 2005). Due to their nonconsensus context, these uORFs are more efficiently translated at low temperature and generally reduce translation efficiency of lFRQ and sFRQ. The mechanism for the temperature-sensitive *frq* transcription is not known.

Further evidence that *frq* can mediate temperature responses in Neurospora comes from a microarray study that identified 14 genes that responded to a 12-h/12-h temperature cycle. All of these temperature-regulated genes were also shown to be clock-controlled and under control of the *frq* locus (Nowrousian *et al.*, 2003). In *frq* null strains, their temperature responses are lost, suggesting that *frq* is required for temperature-dependent regulation of these *ccgs*. Whereas these findings are consistent with a role of FRQ as a temperature sensor, the loss of a temperature response in *frq* null mutants could be a consequence of a nonfunctional clock rather than lack of temperature responsiveness. Because we know that conidiation in *frq-less* strains responds to rhythmic temperature changes (see below), FRQ cannot be the only component required for mediating temperature responses. The relatively low number of genes on the microarray and sensitivity of the technique may provide an explanation for the failure to identify temperature-responsive genes in the *frq* null strain. Alternatively, it may be that most temperature responses are regulated posttranscriptionally.

2. Temperature resetting

As mentioned earlier, *frq* mRNA levels transiently increase or decrease in temperature step-up or step-down experiment (Liu, 2003; Liu *et al.*, 1998). As with light, the acute temperature-induced changes in *frq* and FRQ levels may mediate temperature resetting of the Neurospora clock. This is partly true; however, an investigation of *frq* RNA and protein levels at different temperatures and during step changes in temperature also suggest differences between the mechanisms of clock resetting by light and temperature. At different temperatures, *frq* transcript levels are similar at both high and low temperatures, whereas FRQ protein levels oscillate at higher levels at high temperatures, such that trough levels of FRQ at high temperature are higher than peak levels in FRQ at low temperature. Thus, even in the absence of immediate changes in clock components, a shift to higher temperatures could be interpreted by the clock as starting from the low point in the FRQ oscillation (characteristic of subjective dawn), whereas a shift to lower temperatures starts off from the high point in the FRQ oscillation (characteristic of subjective dusk) (Liu *et al.*, 1998).

3. Temperature limits for circadian rhythmicity and temperature compensation

The temperature-dependent expression of lFRQ and sFRQ provides a mechanism that can explain how temperature limits for the circadian clock are established (Liu *et al.*, 1997). Although either form of FRQ is able to support rhythmicity, both forms are required to generate normal high-amplitude rhythms. Strains with mutations in the first ATG which are only capable of making sFRQ lose rhythmicity at higher temperatures, whereas strains in which the third ATG is mutated and only produce lFRQ become arrhythmic at low temperatures. These data suggest that the two FRQ forms function differently at different temperatures, a notion that is supported by the significantly different period length of strains expressing only one form of FRQ. On the other hand, either form of FRQ can rescue rhythmicity across the entire temperature range when expressed at sufficiently high levels (Colot *et al.*, 2005; Diernfellner *et al.*, 2005; Liu *et al.*, 1997), suggesting that the both quantity and quality of FRQ proteins are important for maintaining robust circadian rhythms in a wild-type strain.

4. Temperature compensation

The discovery of temperature-dependent changes in levels and forms of FRQ protein sparked speculation on whether these changes could be the basis for temperature compensation in Neurospora. In particular, the pronounced temperature dependence in the production of lFRQ and sFRQ could be seen as a possible mechanism by which the feedback cycle could be kept running at a similar pace in different temperatures. Chiefly, the activity of sFRQ would set the speed of the cycle at lower temperatures, gradually lose its efficiency with increasing temperatures under which lFRQ would take over its function with similar efficiency. If this assumption is true, there should be a marked loss in temperature compensation if any of the main ATGs of FRQ are mutated such that only sFRQ or lFRQ can be produced. Such experiments have been independently performed, but with differing outcomes (Colot et al., 2005; Diernfellner et al., 2005; Liu et al., 1997). Whereas in two studies, normal temperature compensation was seen in strains in which sFRQ is ablated (Colot et al., 2005; Liu et al., 1997), another study reports changes in temperature compensation in strains lacking sFRQ (Diernfellner et al., 2005). The reason for this discrepancy is unclear but may be due to the different mutations that were created to abolish sFRQ function. Surprisingly, the levels of sFRQ are the least temperature dependent and the change in ratio between lFRQ and sFRQ is largely due to changes in lFRQ levels at different temperatures. Furthermore, the rhythm in mutants expressing only lFRQ, albeit overcompensated (i.e., runs slower with increasing temperature), has a similar slope of temperature compensation as the wild type (Diernfellner et al., 2005). Both observations suggest that temperature compensation is not simply based on the temperature-dependent expression of two FRQ forms. It is also possible that the interaction and activities of the FFC and the WCC are a function of temperature and the study of their temperature regulation may shed more light on the molecular principles of temperature compensation.

V. TEMPORAL OUTPUT FROM THE OSCILLATOR

Subtractive hybridization, differential screening techniques, and, more recently, microarray experiments have provided a wealth of information on genes whose gene products are regulated by the clock. To date, about 180 ccgs have been identified in independent studies that cover between 10 and 14% of the predicted Neurospora transcriptome. Depending on the study, this translates into 5–20% of the genes investigated, and extrapolating from these data one can expect 500–2000 of the estimated 10,000–11,000 predicted Neurospora genes to be ccgs regulated at the level of transcript abundance. The majority of the ccgs identified so far are influenced by the allelic state of the frequency locus, that is they are arrhythmic in frq-less strains or adopt the periodicity characteristic for the frq period mutant in which they are studied (Bell-Pedersen et al., 1996b; Correa et al., 2003; Lewis et al., 2002; Loros et al., 1989; Nowrousian et al., 2003). A subset of the ccgs cycle in strains lacking a functional frq gene and may be under the control (or be part) of other circadian or noncircadian oscillatory circuits (Correa et al., 2003). Overall, and judging from the predicted or known functions of the identified ccgs, the circadian clock regulates a wide range of functions such as cell division, signaling, development, metabolism, and protein turnover. The influence of Neurospora's clock on such a broad range of processes is similar to the impact of circadian clocks in other organisms (Dunlap and Loros, 2004). However, whether rhythmicity of these ccgs extends to the protein level is in most cases not known.

Considering the large number of ccgs that have been identified, we know relatively little about how the circadian clock, or other oscillators, imposes rhythmic regulation on these targets. Run-on experiments have shown that ccg-1 and eas (=ccg-2) are predominantly regulated at the transcriptional level, and promoter elements have been defined that are necessary for their rhythmic expression (Bell-Pedersen et al., 1996a, 2001). The clock-controlled elements in the ccg promoters differ from the WCC-binding sites identified in the frq promoter. Thus, other clock-controlled transcriptional activators or repressors may relay the clock signal to these two output genes, a conclusion supported by the discovery of clock-controlled transcription factors (Bell-Pedersen et al., 2001; Correa et al., 2003; Lewis et al., 2002). On the other hand, the FRQ and WC complexes may directly regulate promoters of certain output targets and drive their rhythmicity. For example, the promoters of the vvd and al-3 genes contain LREs that are bound by the WCC and may confer rhythmic expression (Bell-Pedersen et al., 2005; Carattoli et al., 1994). In addition, cDNA microarray experiments using a wc-1 mutant strain in which WC-1 is inducibly expressed in the dark has led to the conclusion that 7% of the genes examined are responsive to WC-1 expression (Lewis et al., 2002). Considering the predicted or known functions of some ccgs, for example those involved in protein synthesis/turnover or transcript maturation, one can envision how the circadian clock may influence gene expression at multiple levels and perhaps create a hierarchical system of timed output pathways. A hierarchical organization of circadian systems has been proposed before and in other organisms evidence suggests that subordinate oscillators control a subset of clock-controlled events (Heintzen et al., 1997; Pittendrigh, 1960; Staiger et al., 2003). In addition, it is likely that some clock-controlled output pathways feed back to regulate the central oscillator. The vvd gene is an example of a clock-controlled gene that feeds back to regulate light input into the Neurospora circadian oscillator (Heintzen et al., 2001). Similarly, the expression of the qrf transcript from the frq locus may be considered an output that feeds back to influence light resetting of the clock (Kramer et al., 2003).

To shed more light on how the circadian clock controls its output targets, Vitalini et al. (2004) developed a genetic selection for mutations that affect the regulation of ccg-1 and ccg-2 in a frq-less background in which both genes are aberrantly expressed. Two groups of mutants were identified in which single-gene mutations caused alterations in either ccg-1 levels alone or both ccg-1 and *ccg-2* levels. Because these mutations do not affect the central oscillator, the data are consistent with a model in which the clock indirectly regulates a bifurcated pathway, perhaps via control of a transcription factor required for the expression of both a repressor and an activator of *ccg-1* and *ccg-2*. Mutations that only affect *ccg-1* levels may then lie downstream of such a bifurcation.

VI. FRQ-LESS OSCILLATORS

A number of rhythmic phenomena with periods ranging from a few to over 100 h persist or emerge in the absence of the FRQ-WCC oscillator (Aronson *et al.*, 1994b; Christensen *et al.*, 2004; Correa *et al.*, 2003; Granshaw *et al.*, 2003; He *et al.*, 2005a; Lakin-Thomas and Brody, 2000; Loros and Feldman, 1986; Merrow *et al.*, 1999). Because such rhythms were first noticed in strains lacking a functional *frq* gene (Loros and Feldman, 1986), the oscillators that presumably control these rhythms were given the name FLOs (Iwasaki and Dunlap, 2000). Most of these rhythms are noncircadian, that have highly variable period length, cannot be entrained by light, and their periods are not temperature or nutritionally compensated.

So far, no molecular components of any of the named FLOs have been unambiguously identified. Nitrate reductase activity (NRA) oscillates on a circadian timescale in strains lacking functional FRQ or WC-1 protein and it was proposed that the genes, enzymes, and products of the nitrate assimilation pathway itself may form an autonomous negative feedback loop that rhythmically regulates NRA (Christensen *et al.*, 2004). However, whether this NR-FLO feedback cycle is the basis of rhythmic NRA has not been tested. Conceivably, instead of being an FLO itself, rhythmic NRA may be the target of another FLO. In any case, the discovery of rhythmic NRA provides a potentially useful entry to the molecular dissection of this oscillator. In addition, three recently discovered genes, whose RNA levels display daily rhythms in abundance in a *frq*-less strain, promise to be powerful tools in the dissection of the FLO(s) involved, independent of whether the identified genes are components of the FLO or its output targets (Correa *et al.*, 2003). Perhaps, one of the identified genes, encoding CPC-1, a bZIP transcription factor may have the properties and regulatory potential to be more central to this FLO.

Each of the two FLOs mentioned above produce close to 24-h rhythms but have not been tested for other defining characteristics of true circadian rhythms. In terms of testing its circadian properties, the most rigorously studied FLO and in fact the first one to be described, is manifested as a residual sporulation rhythm in constant darkness that appears in strains lacking functional FRQ protein. Despite being described almost 20 years ago, its molecular components are still unknown and it remains defined by the absence of the FRQ oscillator rather than by a knowledge of its constituents. This lack in progress can, in part, be attributed to the difficulty in taming this highly unpredictable and variable rhythm. A step forward in tackling this problem promised to be the observation that conidiation rhythms in *frq*-less strains can be stably and predictably synchronized to temperature cycles (Merrow *et al.*, 1999). However, in the absence of an independent marker for rhythmicity, it is unclear whether this temperature-controlled FLO (TC-FLO) is at all related to the FLO that free-runs in constant darkness, and in fact, it may be an unrelated phenomenon. Controversy exists regarding the nature of the TC-FLO. Some interpret the characteristics of rhythmic conidiation in temperature cycles as evidence for circadian entrainment (Merrow *et al.*, 1999), whereas others see nothing more than a temperature-driven phenomenon, which terminates in the absence of external temperature cycles (Pregueiro *et al.*, 2005). For example, proponents of the circadian oscillator-theory quote systematic phase angle changes as seen in different symmetrical temperature cycles as hallmarks of circadian rhythmicity (Merrow *et al.*, 1999; Roenneberg *et al.*, 2005), but others find that such changes are dependent on the choice of phase-reference point and that the most commonly used reference points, such as peaks and troughs of conidiation, show no systematic changes (Pregueiro *et al.*, 2005). Other discrepancies exist, such as evidence for frequency demultiplication, another hallmark of an entrained (rather than passively driven) oscillator, which has been observed by some (Roenneberg *et al.*, 2005) but not by others (Pregueiro *et al.*, 2005). In a nutshell, a multitude of factors has contributed to the controversy surrounding the character of the FLO, starting from differences in experimental setups to the final data analysis. The controversy revolves around the nature of the TC-FLO, that is whether it constitutes another circadian oscillator or is a noncircadian, perhaps hourglass-type phenomenon. Agreement exists, however, in the observation that rhythmic conidiation of this FLO stops on releasing Neurospora into constant free-running conditions. In this respect, the TC-FLO is missing a key characteristic of a true circadian rhythm. It is clear that this issue cannot be resolved by race tube analysis alone. Concrete molecular evidence, that is the unambiguous identification of components of this FLO, is required.

Another set of extensively studied FLOs emerge in lipid synthesis mutants, such as *cel* or *chol-1*, and can be manipulated by fatty acid substitution to the growth medium or, become visible on addition of the steroid synthesis intermediates farnesol or geraniol (Granshaw *et al.*, 2003; Lakin-Thomas and Brody, 2000; Mattern and Brody, 1979). The FLOs associated with the *cel* or *chol-1* mutants generally produce very long-period rhythms that can override the FRQ-WC oscillator (FWO) signal, producing rhythms in conidiation that vary between 33 and over 100 h, depending on nutritional conditions and genetic background. On the other hand, the FLO that materializes in the presence of farnesol or geraniol is not seen in the wild type and hence seems to be normally dominated by the FWO (see Dunlap and Loros, 2005; Lakin-Thomas and Brody, 2004 for a more detailed discussion of these FLOs).

Despite the undoubted existence of FLOs, there is little evidence to suggest that they contribute to robust circadian rhythmicity as seen in Neurospora wild-type strains. Many factors, including lipid content, carbon source, temperature, and growth conditions, can significantly impact on the period or even the existence of some of the FLOs, while they generally have little or no effects on the FWO function in a wild-type strain. It is possible, however, that in the wild-type, these FLOs couple to the FWO and work downstream to control rhythmic outputs in Neurospora (Dunlap et al., 2004; Lakin-Thomas and Brody, 2004). In summary, little progress has been made in identifying the nature, let alone the molecular components of the elusive FLOs. This aside, their intense study and associated controversy has raised awareness and rekindled interest in other rhythmic phenomena, and their potential interactions with circadian oscillators (Pregueiro et al., 2005; Roenneberg et al., 2005).

VII. CONCLUSIONS

Over the last few years, much progress has been made in our understanding of the molecular basis of circadian rhythmicity. In Neurospora, the main photoreceptors that feed light information into the circadian clock have been identified and our picture of the molecular principles of temperature sensing has become clearer. Many important clock components that help to maintain robust circadian rhythmicity have been identified. The identification of FRH as a core component of the circadian negative feedback loop raises the possibility of clock-controlled RNA metabolism. The identification of kinases and phosphatases that use clock proteins as their substrate and the study of protein degradation pathways has led to an appreciation of the importance of clock protein turnover for maintaining the circadian feedback cycle. These recent advances in our understanding of negative feedback regulation in Neurospora have made its circadian oscillator one of the best understood among eukaryotes. In addition, our understanding of the pathways that are important for light–dark entrainment of the circadian clock is rapidly improving. At the same time, the number of circadian ccgs has risen dramatically, leaving much to do in terms of understanding their functions within and their importance for the circadian system. In contrast, little progress has been made in identifying the molecular components of oscillators that operate outside or in the absence of the known FRQ-WC oscillator. However, the recent identification of components involved in nitrate utilization and genes that are rhythmically expressed in the absence of a functional FWO may provide a first molecular entry into these otherwise elusive oscillators.

Generally, it is thought that eukaryotic circadian systems are more complex than our current understanding of them reveals and probably involve a multitude of regulatory networks in which some oscillatory feedback loops will take the lead over or interact with other circadian and noncircadian oscillators (Bell-Pedersen *et al.*, 2005). Although the assumption of complexity is always merited, it is, however, not a necessary feature of circadian systems, at least with respect to the number of components and regulatory principles required. This was recently demonstrated for the cyanobacterial clock, in which, and independent of transcription and translation, only three clock proteins (KaiA, KaiB, and KaiC) and ATP are required to generate temperature compensated, circadian rhythms in KaiC phosphorylation (Nakajima *et al.*, 2005; Tomita *et al.*, 2005). It is possible that we underestimate the complexity of the minimal cyanobacterial system but it seems also prudent not to assume complexity of eukaryotic clocks beyond necessity. This aside, cyanobacterial and eukaryotic clocks seem fundamentally different and this may extend to the level of complexity of the circadian network (Golden, 2003). In this respect, the simple eukaryote Neurospora may become one of the best and most completely understood eukaryotic model organisms for circadian and noncircadian phenomena.

Acknowledgments

We would like to thank Sue Crosthwaite, Mark Elvin, Suzanne Hunt, and members of the Liu laboratory for critical reading of the chapter. This work was supported by grants from the BBSRC (34/G17511 to C.H.) and National Institutes of Health (GM062591 and GM068496 to Y.L.) and Welch Foundation (to Y.L.). Y.L. is the Louise W. Kahn endowed scholar in Biomedical Research at University of Texas Southwestern Medical Center.

References

Akten, B., Jauch, E., Genova, G. K., Kim, E. Y., Edery, I., Raabe, T., and Jackson, F. R. (2003). A role for CK2 in the *Drosophila* circadian oscillator. *Nat. Neurosci.* **6**, 251–257.

Allmang, C., Kufel, J., Chanfreau, G., Mitchell, P., Petfalski, E., and Tollervey, D. (1999). Functions of the exosome in rRNA, snoRNA and snRNA synthesis. *EMBO J.* **18**, 5399–5410.

Aronson, B. D., Johnson, K. A., Loros, J. J., and Dunlap, J. C. (1994a). Negative feedback defining a circadian clock: Autoregulation in the clock gene *frequency*. *Science* **263**, 1578–1584.

Aronson, B. D., Johnson, K. A., and Dunlap, J. C. (1994b). The circadian clock locus *frequency*: A single ORF defines period length and temperature compensation. *Proc. Natl. Acad. Sci. USA* **91**, 7683–7687.

Ashmore, L. J., and Sehgal, A. (2003). A fly's eye view of circadian entrainment. *J. Biol. Rhythms* **18**, 206–216.

Ballario, P., Vittorioso, P., Magrelli, A., Talora, C., Cabibbo, A., and Macino, G. (1996). *White collar-1*, a central regulator of blue-light responses in *Neurospora crassa*, is a zinc-finger protein. *EMBO J.* **15**, 1650–1657.

Ballario, P., Talora, C., Galli, D., Linden, H., and Macino, G. (1998). Roles in dimerization and blue light photoresponses of PAS and LOV domains of *Neurospora crassa* white collar proteins. *Mol. Microbiol.* **29,** 719–729.

Bell-Pedersen, D., Dunlap, J. C., and Loros, J. J. (1996a). Distinct cis-acting elements mediate clock, light, and developmental regulation of the *Neurospora crassa eas(ccg-2)* gene. *Mol. Cell. Biol.* **16,** 513–521.

Bell-Pedersen, D., Shinohara, M., Loros, J., and Dunlap, J. C. (1996b). Circadian clock-controlled genes isolated from *Neurospora crassa* are late night to early morning specific. *Proc. Natl. Acad. Sci. USA* **93,** 13096–13101.

Bell-Pedersen, D., Lewis, Z. A., Loros, J. J., and Dunlap, J. C. (2001). The *Neurospora* circadian clock regulates a transcription factor that controls rhythmic expression of the output *eas(ccg-2)* gene. *Mol. Microbiol.* **41,** 897–909.

Bell-Pedersen, D., Cassone, V. M., Earnest, D. J., Golden, S. S., Hardin, P. E., Thomas, T. L., and Zoran, M. J. (2005). Circadian rhythms from multiple oscillators: Lessons from diverse organisms. *Nat. Rev. Genet.* **6,** 544–556.

Borkovich, K. A., Alex, L. A., Yarden, O., Freitag, M., Turner, G. E., Read, N. D., Seiler, S., Bell-Pedersen, D., Paietta, J., Plesofsky, N., Plamann, M., Goodrich-Tanrikulu, *et al.* (2004). Lessons from the genome sequence of *Neurospora crassa*: Tracing the path from genomic blueprint to multicellular organism. *Microbiol. Mol. Biol. Rev.* **68,** 1–108.

Brown, S. A., Ripperger, J., Kadener, S., Fleury-Olela, F., Vilbois, F., Rosbash, M., and Schibler, U. (2005). PERIOD1-associated proteins modulate the negative limb of the mammalian circadian oscillator. *Science* **308,** 693–696.

Bünning, E. (1936). Die endogene Tagesrhythmik als Grundlage der photoperiodischen Reaktion. *Ber. Dtsch. Bot. Ges.* **54,** 590–607.

Carattoli, A., Cogoni, C., Morelli, G., and Macino, G. (1994). Molecular characterization of upstream regulatory sequences controlling the photoinduced expression of the *albino-3* gene of *Neurospora crassa*. *Mol. Microbiol.* **13,** 787–795.

Cheng, P., Yang, Y., Heintzen, C., and Liu, Y. (2001a). Coiled-coil domain mediated FRQ-FRQ interaction is essential for its circadian clock function in *Neurospora*. *EMBO J.* **20,** 101–108.

Cheng, P., Yang, Y., and Liu, Y. (2001b). Interlocked feedback loops contribute to the robustness of the Neurospora circadian clock. *Proc. Natl. Acad. Sci. USA* **98,** 7408–7413.

Cheng, P., Yang, Y., Gardner, K. H., and Liu, Y. (2002). PAS domain-mediated WC-1/WC-2 interaction is essential for maintaining the steady state level of WC-1 and the function of both proteins in circadian clock and light responses of *Neurospora*. *Mol. Cell. Biol.* **22,** 517–524.

Cheng, P., He, Q., Yang, Y., Wang, L., and Liu, Y. (2003a). Functional conservation of light, oxygen, or voltage domains in light sensing. *Proc. Natl. Acad. Sci. USA* **100,** 5938–5943.

Cheng, P., Yang, Y., Wang, L., He, Q., and Liu, Y. (2003b). WHITE COLLAR-1, a multifunctional *Neurospora* protein involved in the circadian feedback loops, light sensing, and transcription repression of *wc-2*. *J. Biol. Chem.* **278,** 3801–3808.

Cheng, P., He, Q., He, Q., Wang, L., and Liu, Y. (2005). Regulation of the *Neurospora* circadian clock by an RNA helicase. *Genes Dev.* **19,** 234–241.

Christensen, M. K., Falkeid, G., Loros, J. J., Dunlap, J. C., Lillo, C., and Ruoff, P. (2004). A nitrate-induced *frq*-less oscillator in *Neurospora crassa*. *J. Biol. Rhythms* **19,** 280–286.

Collett, M. A., Garceau, N., Dunlap, J. C., and Loros, J. J. (2002). Light and clock expression of the *Neurospora* clock gene *frequency* is differentially driven by but dependent on WHITE COLLAR-2. *Genetics* **160,** 149–158.

Colot, H. V., Loros, J. J., and Dunlap, J. C. (2005). Temperature-modulated alternative splicing and promoter use in the circadian clock gene *frequency*. *Mol. Biol. Cell* **16,** 5563–5571.

Colot, H. V., Park, G., Turner, G. E., Ringelberg, C., Crew, C., Litvinkova, L., Weiss, R. L., Borkovich, K. A., and Dunlap, J. C. (2006). A high throughput gene knockout procedure for

Neurospora reveals functions for multiple transcription factors. *Proc. Natl. Acad. Sci. USA* **103,** 10352–10357.

Correa, A., Lewis, A. Z., Greene, A. V., March, I. J., Gomer, R. H., and Bell-Pedersen, D. (2003). Multiple oscillators regulate circadian gene expression in *Neurospora. Proc. Natl. Acad. Sci. USA* **100,** 13597–13602.

Crosson, S., and Moffat, K. (2001). Structure of a flavin-binding plant photoreceptor domain: Insights into light-mediated signal transduction. *Proc. Natl. Acad. Sci. USA* **98,** 2995–3000.

Crosson, S., and Moffat, K. (2002). Photoexcited structure of a plant photoreceptor domain reveals a light-driven molecular switch. *Plant Cell* **14,** 1067–1075.

Crosson, S., Rajagopal, S., and Moffat, K. (2003). The LOV domain family: Photoresponsive signaling modules coupled to diverse output domains. *Biochemistry* **42,** 2–10.

Crosthwaite, S. K. (2004). Circadian clocks and natural antisense RNA. *FEBS Lett.* **567,** 49–54.

Crosthwaite, S. K., Loros, J. J., and Dunlap, J. C. (1995). Light-induced resetting of a circadian clock is mediated by a rapid increase in *frequency* transcript. *Cell* **81,** 1003–1012.

Crosthwaite, S. K., Dunlap, J. C., and Loros, J. J. (1997). Neurospora *wc-1* and *wc-2*: Transcription, photoresponses, and the origins of circadian rhythmicity. *Science* **276,** 763–769.

de la Cruz, J., Kressler, D., Tollervey, D., and Linder, P. (1998). Dob1p (Mtr4p) is a putative ATP-dependent RNA helicase required for the 3′ end formation of 5.8S rRNA in *Saccharomyces cerevisiae. EMBO J.* **17,** 1128–1140.

Denault, D. L., Loros, J. J., and Dunlap, J. C. (2001). WC-2 mediates WC-1–FRQ interaction within the PAS protein-linked circadian feedback loop of *Neurospora. EMBO J.* **20,** 109–117.

Diernfellner, A. C., Schafmeier, T., Merrow, M. W., and Brunner, M. (2005). Molecular mechanism of temperature sensing by the circadian clock of *Neurospora crassa. Genes Dev.* **19,** 1968–1973.

Dragovic, Z., Tan, Y., Gorl, M., Roenneberg, T., and Merrow, M. (2002). Light reception and circadian behavior in 'blind' and 'clock-less' mutants of *Neurospora crassa. EMBO J.* **21,** 3643–3651.

Dunlap, J. C. (1999). Molecular bases for circadian clocks. *Cell* **96,** 271–290.

Dunlap, J. C., and Loros, J. J. (2004). The *Neurospora* circadian system. *J. Biol. Rhythms* **19,** 414–424.

Dunlap, J. C., and Loros, J. J. (2005). Analysis of circadian rhythms in *Neurospora*: Overview of assays and genetic and molecular biological manipulation. *Methods Enzymol.* **393,** 3–22.

Dunlap, J. C., Loros, J. J., and DeCoursey, P. J. (2004). "Chronobiology." Sinauer Associates Inc., Publishers, Sunderland.

Eide, E. J., Woolf, M. F., Kang, H., Woolf, P., Hurst, W., Camacho, F., Vielhaber, E. L., Giovanni, A., and Virshup, D. M. (2005). Control of mammalian circadian rhythm by CKIepsilon-regulated proteasome-mediated PER2 degradation. *Mol. Cell. Biol.* **25,** 2795–2807.

Elvin, M., Loros, J., Dunlap, J. C., and Heintzen, C. (2005). The PAS/LOV protein VIVID supports a rapidly dampened daytime oscillator that facilitates entrainment of the *Neurospora* circadian clock. *Genes Dev.* **19,** 2593–2605.

Feldman, J. F., and Hoyle, M. (1973). Isolation of circadian clock mutants of *Neurospora crassa. Genetics* **75,** 605–613.

Fleissner, G., and Fleissner, G. (1992). Feedback loops in the circadian system. *In* "Circadian Rhythms" (M. Zatz, ed.), pp. 79–84. Elsevier, Amsterdam.

Franchi, L., Fulci, V., and Macino, G. (2005). Protein kinase C modulates light responses in *Neurospora* by regulating the blue light photoreceptor WC-1. *Mol. Microbiol.* **56,** 334–345.

Froehlich, A. C., Liu, Y., Loros, J. J., and Dunlap, J. C. (2002). WHITE COLLAR-1, a circadian blue light photoreceptor, binding to the *frequency* promoter. *Science* **297,** 815–819.

Froehlich, A. C., Loros, J. J., and Dunlap, J. C. (2003). Rhythmic binding of a WHITE COLLAR-containing complex to the *frequency* promoter is inhibited by FREQUENCY. *Proc. Natl. Acad. Sci. USA* **100,** 5914–5919.

Froehlich, A. C., Noh, B., Vierstra, R. D., Loros, J. J., and Dunlap, J. C. (2005). Genetic and molecular analysis of phytochromes from the filamentous fungus *Neurospora crassa*. *Eukaryotic Cell* **4,** 2140–2152.

Galagan, J. E., Calvo, S. E., Borkovich, K. A., Selker, E. U., Read, N. D., Jaffe, D., FitzHugh, W., Ma, L. J., Smirnov, S., Purcell, S., Rehman, B., Elkins, T., *et al.* (2003). The genome sequence of the filamentous fungus *Neurospora crassa. Nature* **422,** 859–868.

Garceau, N., Liu, Y., Loros, J. J., and Dunlap, J. C. (1997). Alternative initiation of translation and time-specific phosphorylation yield multiple forms of the essential clock protein FREQUENCY. *Cell* **89,** 469–476.

Gehring, W., and Rosbash, M. (2003). The coevolution of blue-light photoreception and circadian rhythms. *J. Mol. Evol.* **57,** S286–S289.

Gillette, M. U., and Mitchell, J. W. (2002). Signaling in the suprachiasmatic nucleus: Selectively responsive and integrative. *Cell Tissue Res.* **309,** 99–107.

Glossop, N. R., Lyons, L. C., and Hardin, P. E. (1999). Interlocked feedback loops within the *Drosophila* circadian oscillator. *Science* **286,** 766–768.

Golden, S. S. (2003). Timekeeping in bacteria: The cyanobacterial circadian clock. *Curr. Opin. Microbiol.* **6,** 535–540.

Gooch, V. D. (1985). Effects of light and temperature steps on circadian rhythms of *Neurospora crassa* and *Gonyaulax polyedra. In* "Temporal Order" (N. I. Jaeger, ed.), pp. 232–237. Springer-Verlag, New York.

Gooch, V. D., Freeman, L., and Lakin-Thomas, P. L. (2004). Time-lapse analysis of the circadian rhythms of conidiation and growth rate in *Neurospora. J. Biol. Rhythms* **19,** 493–503.

Gorl, M., Merrow, M., Huttner, B., Johnson, J., Roenneberg, T., and Brunner, M. (2001). A PEST-like element in FREQUENCY determines the length of the circadian period in *Neurospora crassa. EMBO J.* **20,** 7074–7084.

Granshaw, T., Tsukamoto, M., and Brody, S. (2003). Circadian rhythms in *Neurospora crassa*: Farnesol or geraniol allow expression of rhythmicity in the otherwise arrhythmic strains frq^{10}, *wc-1*, and *wc-2. J. Biol. Rhythms* **18,** 287–296.

Grima, B., Lamouroux, A., Chelot, E., Papin, C., Limbourg-Bouchon, B., and Rouyer, F. (2002). The F-box protein slimb controls the levels of clock proteins period and timeless. *Nature* **420,** 178–182.

Harper, S. M., Neil, L. C., and Gardner, K. H. (2003). Structural basis of a phototropin light switch. *Science* **301,** 1541–1544.

He, Q., and Liu, Y. (2005). Molecular mechanism of light responses in *Neurospora*: From light-induced transcription to photoadaptation. *Genes Dev.* **19,** 2888–2899.

He, Q., Cheng, P., Yang, Y., Wang, L., Gardner, K. H., and Liu, Y. (2002). WHITE COLLAR-1, a DNA binding transcription factor and a light sensor. *Science* **297,** 840–843.

He, Q., Cheng, P., Yang, Y., He, Q., Yu, H., and Liu, Y. (2003). FWD1-mediated degradation of FREQUENCY in *Neurospora* establishes a conserved mechanism for circadian clock regulation. *EMBO J.* **22,** 4421–4430.

He, Q., Cheng, P., He, Q. Y., and Liu, Y. (2005a). The COP9 signalosome regulates the *Neurospora* circadian clock by controlling the stability of the SCFFWD-1 complex. *Genes Dev.* **19,** 1518–1531.

He, Q., Shu, H., Cheng, P., Chen, S., Wang, L., and Liu, Y. (2005b). Light-independent phosphorylation of WHITE COLLAR-1 regulates its function in the *Neurospora* circadian negative feedback loop. *J. Biol. Chem.* **280,** 17526–17532.

He, Q., Cha, J., Lee, H., Yang, Y., and Liu, Y. (2006). CKI and CKII mediate the FREQUENCY-dependent phosphorylation of the WHITE COLLAR complex to close the *Neurospora* circadian negative feedback loop. *Genes & Dev.* **20,** 2252–2565.

Heintzen, C. (2005). PAS proteins and the *Neurospora* circadian clock. *In* "The Circadian Clock in Eukaryotic Microbes" (F. Kippert, ed.). Landes/Eurekah Bioscience, Austin.

Heintzen, C., Nater, M., Apel, K., and Staiger, D. (1997). AtGRP7, a nuclear RNA-binding protein as a component of a circadian-regulated negative feedback loop in *Arabidopsis thaliana. Proc. Natl. Acad. Sci. USA* **94**, 8515–8520.

Heintzen, C., Loros, L. L., and Dunlap, J. C. (2001). The PAS protein VIVID defines a clock-associated feedback loop that represses light input, modulates gating and regulates clock resetting. *Cell* **104**, 453–464.

Hilleren, P. J., and Parker, R. (2003). Cytoplasmic degradation of splice-defective pre-mRNAs and intermediates. *Mol. Cell* **12**, 1453–1465.

Huang, G., Wang, L., and Liu, Y. (2006). Molecular mechanism of suppression of circadian rhythm by a critical stimulus. *EMBO J.* **25**, 5349–5357.

Imaizumi, T., Tran, H. G., Swartz, T. E., Briggs, W. R., and Kay, S. A. (2003). FKF1 is essential for photoperiodic-specific light signalling in Arabidopsis. *Nature* **426**, 302–306.

Imaizumi, T., Schultz, T. F., Harmon, F. G., Ho, L. A., and Kay, S. A. (2005). FKF1F-BOX protein mediates cyclic degradation of a repressor of CONSTANS in *Arabidopsis. Science* **309**, 293–297.

Iwasaki, H., and Dunlap, J. C. (2000). Microbial circadian oscillatory systems in *Neurospora* and *Drosophila*: Models for cellular clocks. *Curr. Opin. Microbiol.* **3**, 189–196.

Jacobs, J. S., Anderson, A. R., and Parker, R. P. (1998). The 3′ to 5′ degradation of yeast mRNAs is a general mechanism for mRNA turnover that requires the SKI2 DEVH box protein and 3′ to 5′ exonucleases of the exosome complex. *EMBO J.* **17**, 1497–1506.

Johnson, C. H., Elliott, J. A., and Foster, R. (2003). Entrainment of circadian programs. *Chronobiol. Int.* **20**, 741–774.

Johnson, C. H., Elliott, J., Foster, R., Honma, K., and Kronauer, R. E. (2004). Fundamental properties of circadian rhythms. In "Chronobiology" (J. C. Dunlap, J. J. Loros, and P. J. DeCoursey, eds.). Sinauer Associates Inc., Publishers, Sunderland.

Kim, E. Y., Bae, K., Ng, F. S., Glossop, N. R., Hardin, P. E., and Edery, I. (2002). *Drosophila* CLOCK protein is under posttranscriptional control and influences light-induced activity. *Neuron* **34**, 69–81.

Kloss, B., Price, J. L., Saez, L., Blau, J., Rothenfluh, A., and Young, M. W. (1998). The *Drosophila* clock gene *double-time* encodes a protein closely related to human casein kinase Iε. *Cell* **94**, 97–107.

Ko, H. W., Jiang, J., and Edery, I. (2002). Role for Slimb in the degradation of *Drosophila* period protein phosphorylated by Doubletime. *Nature* **420**, 673–678.

Kramer, C., Loros, J. J., Dunlap, J. C., and Crosthwaite, S. K. (2003). Role for antisense RNA in regulating circadian clock function in *Neurospora crassa. Nature* **421**, 948–952.

Krishnan, B., Levine, J. D., Lynch, M. K., Dowse, H. B., Funes, P., Hall, J. C., Hardin, P. E., and Dryer, S. E. (2001). A new role for *cryptochrome* in a *Drosophila* circadian oscillator. *Nature* **411**, 313–317.

Lakin-Thomas, P. L., and Brody, S. (2000). Circadian rhythms in *Neurospora crassa*: Lipid deficiencies restore robust rhythmicity to null *frequency* and *white-collar* mutants. *Proc. Natl. Acad. Sci. USA* **97**, 256–261.

Lakin-Thomas, P. L., and Brody, S. (2004). Circadian rhythms in microorganisms: New complexities. *Annu. Rev. Microbiol.* **58**, 489–519.

Lee, K., Loros, J. J., and Dunlap, J. C. (2000). Interconnected feedback loops in the *Neurospora* circadian system. *Science* **289**, 107–110.

Lewis, Z. A., Correa, A., Schwerdtfeger, C., Link, K. L., Xie, X., Gomer, R. H., Thomas, T., Ebbole, D. J., and Bell-Pedersen, D. (2002). Overexpression of WHITE COLLAR-1 (WC-1) activates circadian clock-associated genes, but is not sufficient to induce most light-regulated gene expression in *Neurospora crassa. Mol. Microbiol.* **45**, 917–931.

Lin, J. M., Kilman, V. L., Keegan, K., Paddock, B., Emery-Le, M., Rosbash, M., and Allada, R. (2002). A role for casein kinase 2alpha in the *Drosophila* circadian clock. *Nature* **420**, 816–820.

Linden, H. (2002). Circadian rhythms: A WHITE COLLAR protein senses blue light. *Science* **297**, 777–778.

Linden, H., and Macino, G. (1997). White collar-2, a partner in blue-light signal transduction, controlling expression of light-regulated genes in *Neurospora crassa*. *EMBO J.* **16**, 98–109.

Linden, H., Ballario, P., Arpaia, G., and Macino, G. (1999). Seeing the light: News in *Neurospora* blue light signal transduction. *Adv. Genet.* **37**, 35–54.

Liu, Y. (2003). Molecular mechanisms of entrainment in the *Neurospora* circadian clock. *J. Biol. Rhythms* **18**, 195–205.

Liu, Y. (2005). Analysis of posttranslational regulations in the *Neurospora* circadian clock. *Methods Enzymol.* **393**, 379–393.

Liu, Y., Garceau, N., Loros, J. J., and Dunlap, J. C. (1997). Thermally regulated translational control mediates an aspect of temperature compensation in the *Neurospora* circadian clock. *Cell* **89**, 477–486.

Liu, Y., Merrow, M. M., Loros, J. J., and Dunlap, J. C. (1998). How temperature changes reset a circadian oscillator. *Science* **281**, 825–829.

Liu, Y., Loros, J., and Dunlap, J. C. (2000). Phosphorylation of the *Neurospora* clock protein FREQUENCY determines its degradation rate and strongly influences the period length of the circadian clock. *Proc. Natl. Acad. Sci. USA* **97**, 234–239.

Liu, Y., He, Q., and Cheng, P. (2003). Photoreception in *Neurospora*: A tale of two WHITE COLLAR proteins. *Cell. Mol. Life Sci.* **60**, 2131–2138.

Loros, J. J., and Feldman, J. F. (1986). Loss of temperature compensation of circadian period length in the *frq-9* mutant of *Neurospora crassa*. *J. Biol. Rhythms* **1**, 187–198.

Loros, J. J., and Dunlap, J. C. (2001). Genetic and molecular analysis of circadian rhythms in *Neurospora*. *Annu. Rev. Physiol.* **63**, 757–794.

Loros, J. J., Richman, A., and Feldman, J. F. (1986). A recessive circadian clock mutant at the *frq* locus in *Neurospora crassa*. *Genetics* **114**, 1095–1110.

Loros, J. J., Denome, S. A., and Dunlap, J. C. (1989). Molecular cloning of genes under the control of the circadian clock in *Neurospora*. *Science* **243**, 385–388.

Lowrey, P. L., Shimomura, K., Antoch, M. P., Yamazaki, S., Zemenides, P. D., Ralph, M. R., Menaker, M., and Takahashi, J. S. (2000). Positional syntenic cloning and functional characterization of the mammalian circadian mutation *tau*. *Science* **288**, 483–492.

Luo, C., Loros, J. J., and Dunlap, J. C. (1998). Nuclear localization is required for function of the essential clock protein FREQUENCY. *EMBO J.* **17**, 1228–1235.

Martinek, S., Inonog, S., Manoukian, A. S., and Young, M. W. (2001). A role for the segment polarity gene shaggy/GSK-3 in the *Drosophila* circadian clock. *Cell* **105**, 769–779.

Mattern, D., and Brody, S. (1979). Circadian rhythms in *Neurospora crassa*: Effects of unsaturated fatty acids. *J. Bacteriol.* **139**, 977–988.

McClung, C. R., Fox, B. A., and Dunlap, J. C. (1989). The *Neurospora* clock gene *frequency* shares a sequence element with the *Drosophila* clock gene *period*. *Nature* **339**, 558–562.

McWatters, H. G., Bastow, R. M., Hall, A., and Millar, A. J. (2000). The ELF3 zeitnehmer regulates light signalling to the circadian clock. *Nature* **408**, 716–720.

Merrow, M., Garceau, N., and Dunlap, J. C. (1997). Dissection of a circadian oscillation into discrete domains. *Proc. Natl. Acad. Sci. USA* **94**, 3877–3882.

Merrow, M., Brunner, M., and Roenneberg, T. (1999). Assignment of circadian function for the *Neurospora* clock gene *frequency*. *Nature* **399**, 584–586.

Merrow, M., Franchi, L., Dragovic, Z., Gorl, M., Johnson, J., Brunner, M., Macino, G., and Roenneberg, T. (2001). Circadian regulation of the light input pathway in *Neurospora crassa*. *EMBO J.* **20**, 307–315.

Millar, A. J., and Kay, S. A. (1996). Integration of circadian and phototransduction pathways in the network controlling CAB gene transcription in *Arabidopsis*. *Proc. Natl. Acad. Sci. USA* **93**, 15491–15496.

Mitchell, P., and Tollervey, D. (2000). Musing on the structural organization of the exosome complex. *Nat. Struct. Biol.* **7**, 843–846.

Morgan, L. W., Greene, A. V., and Bell-Pedersen, D. (2003). Circadian and light-induced expression of luciferase in *Neurospora crassa. Fungal Genet. Biol.* **38**, 327–332.

Nakajima, M., Imai, K., Ito, H., Nishiwaki, T., Murayama, Y., Iwasaki, T., Oyama, T., and Kondo, T. (2005). Reconstitution of circadian oscillation of cyanobacterial KaiC phosphorylation *in vitro. Science* **308**, 414–415.

Nawathean, P., and Rosbash, M. (2004). The doubletime and CKII kinases collaborate to potentiate Drosophila PER transcriptional repressor activity. *Mol. Cell* **13**, 213–223.

Nowrousian, M., Duffield, G. E., Loros, J. J., and Dunlap, J. C. (2003). The *frequency* gene is required for temperature-dependent regulation of many clock-controlled genes in *Neurospora crassa. Genetics* **164**, 923–933.

Ouyang, Y., Andersson, C. R., Kondo, T., Golden, S. S., and Johnson, C. H. (1998). Resonating circadian clocks enhance fitness in cyanobacteria. *Proc. Natl. Acad. Sci. USA* **95**, 8660–8664.

Pittendrigh, C. S. (1960). Circadian rhythms and the circadian organization of living things. In "Cold Spring Harbor Symp. Quant. Biol. XXV: Biological Clocks" (A. Chovnick, ed.), pp. 159–184. Cold Spring Harbor Press, New York.

Pregueiro, A. M., Price-Lloyd, N., Bell-Pedersen, D., Heintzen, C., Loros, J. J., and Dunlap, J. C. (2005). Assignment of an essential role for the Neurospora *frequency* gene in circadian entrainment to temperature cycles. *Proc. Natl. Acad. Sci. USA* **102**, 2210–2215.

Preitner, N., Damiola, F., Luis-Lopez, M., Zakany, J., Duboule, D., Albrecht, U., and Schibler, U. (2002). The orphan nuclear receptor REV-ERBalpha controls circadian transcription within the positive limb of the mammalian circadian oscillator. *Cell* **110**, 251–260.

Price, J. L., Blau, J., Rothenfluh, A., Adodeely, M., Kloss, B., and Young, M. W. (1998). *Doubletime* is a new Drosophila clock gene that regulates PERIOD protein accumulation. *Cell* **94**, 83–95.

Roenneberg, T., and Merrow, M. (2001). Seasonality and photoperiodism in fungi. *J. Biol. Rhythms* **16**, 403–414.

Roenneberg, T., Daan, S., and Merrow, M. (2003). The art of entrainment. *J. Biol. Rhythms* **18**, 183–194.

Roenneberg, T., Dragovic, Z., and Merrow, M. (2005). Demasking biological oscillators: Properties and principles of entrainment exemplified by the *Neurospora* circadian clock. *Proc. Natl. Acad. Sci. USA* **102**, 7742–7747.

Sathyanarayanan, S., Zheng, X., Xiao, R., and Sehgal, A. (2004). Posttranslational regulation of *Drosophila* PERIOD protein by protein phosphatase 2A. *Cell* **116**, 603–615.

Saunders, D. S. (2005). Erwin Bünning and Tony Lees, two giants of chronobiology, and the problem of time measurement in insect photoperiodism. *J. Insect Physiol.* **51**, 599–608.

Schafmeier, T., Haase, A., Kaldi, K., Scholz, J., Fuchs, M., and Brunner, M. (2005). Transcriptional feedback of a Neurospora circadian clock gene by phosphorylation-dependent inactivation of its transcription factor. *Cell* **122**, 235–246.

Schwerdtfeger, C., and Linden, H. (2000). Localization and light-dependent phosphorylation of white collar-1 and 2, the two central components of blue light signaling in *Neurospora crassa. Eur. J. Biochem.* **267**, 414–422.

Schwerdtfeger, C., and Linden, H. (2001). Blue light adaptation and desensitization of light signal transduction in *Neurospora crassa. Mol. Microbiol.* **39**, 1080–1087.

Schwerdtfeger, C., and Linden, H. (2003). VIVID is a flavoprotein and serves as a fungal blue light photoreceptor for photoadaptation. *EMBO J.* **22**, 4846–4855.

Shearman, L. P., Sriram, S., Weaver, D. R., Maywood, E. S., Chaves, I., Zheng, B., Kume, K., Lee, C. C., van der Horst, G. T., Hastings, M. H., and Reppert, S. M. (2000). Interacting molecular loops in the mammalian circadian clock. *Science* **288**, 1013–1019.

Shigeyoshi, Y., Taguchi, K., Yamamoto, S., Takeida, S., Yan, L., Tei, H., Moriya, S., Shibata, S., Loros, J. J., Dunlap, J. C., and Okamura, H. (1997). Light-induced resetting of a mammalian circadian clock is associated with rapid induction of the *mPer1* transcript. *Cell* **91**, 1043–1053.

Shrode, L. B., Lewis, Z. A., White, L. D., Bell-Pedersen, D., and Ebbole, D. J. (2001). *vvd* is required for light adaptation of conidiation-specific genes of *Neurospora crassa*, but not circadian conidiation. *Fungal Genet. Biol.* **32**, 169–181.

Staiger, D., Zecca, L., Kirk, D. A. W., Apel, K., and Eckstein, L. (2003). The circadian clock regulated RNA-binding protein AtGRP7 autoregulates its expression by influencing alternative splicing of its own pre-mRNA. *Plant J.* **33**, 361–371.

Suarez-Lopez, P., Wheatley, K., Robson, F., Onouchi, H., Valverde, F., and Coupland, G. (2001). CONSTANS mediates between the circadian clock and the control of flowering in Arabidopsis. *Nature* **410**, 1116–1120.

Talora, C., Franchi, L., Linden, H., Ballario, P., and Macino, G. (1999). Role of a WHITE COLLAR-1:WHITE COLLAR-2 complex in blue-light signal transduction. *EMBO J.* **18**, 4961–4968.

Tan, Y., Dragovic, Z., Roenneberg, T., and Merrow, M. (2004a). Entrainment dissociates transcription and translation of a circadian clock gene in *Neurospora*. *Curr. Biol.* **14**, 433–438.

Tan, Y., Merrow, M., and Roenneberg, T. (2004b). Photoperiodism in *Neurospora crassa*. *J. Biol. Rhythms* **19**, 135–143.

Tauber, E., Last, K. S., Olive, P. J., and Kyriacou, C. P. (2004). Clock gene evolution and functional divergence. *J. Biol. Rhythms* **19**, 445–458.

Tomita, J., Nakajima, M., Kondo, T., and Iwasaki, H. (2005). No transcription-translation feedback in circadian rhythm of KaiC phosphorylation. *Science* **307**, 251–254.

Torchet, C., Bousquet-Antonelli, C., Milligan, L., Thompson, E., Kufel, J., and Tollervey, D. (2002). Processing of 3′-extended read-through transcripts by the exosome can generate functional mRNAs. *Mol. Cell* **9**, 1285–1296.

Valverde, F., Mouradov, A., Soppe, W., Ravenscroft, D., Samach, A., and Coupland, G. (2004). Photoreceptor regulation of CONSTANS protein in photoperiodic flowering. *Science* **303**, 1003–1006.

Vitalini, M. W., Morgan, L. W., March, I. J., and Bell-Pedersen, D. (2004). A genetic selection for circadian output pathway mutations in *Neurospora crassa*. *Genetics* **167**, 119–129.

Xu, Y., Padiath, Q. S., Shapiro, R. E., Jones, C. R., Wu, S. C., Saigoh, N., Saigoh, K., Ptacek, L. J., and Fu, Y. H. (2005). Functional consequences of a *CKIdelta* mutation causing familial advanced sleep phase syndrome. *Nature* **434**, 640–644.

Yang, Y., Cheng, P., Zhi, G., and Liu, Y. (2001). Identification of a calcium/calmodulin-dependent protein kinase that phosphorylates the *Neurospora* circadian clock protein FREQUENCY. *J. Biol. Chem.* **276**, 41064–41072.

Yang, Y., Cheng, P., and Liu, Y. (2002). Regulation of the *Neurospora* circadian clock by CASEIN KINASE II. *Genes Dev.* **16**, 994–1006.

Yang, Y., Cheng, P., He, Q., Wang, L., and Liu, Y. (2003). Phosphorylation of FREQUENCY protein by CASEIN KINASE II is necessary for the function of the *Neurospora* circadian clock. *Mol. Cell. Biol.* **23**, 6221–6228.

Yang, Y., He, Q., Cheng, P., Wrage, P., Yarden, O., and Liu, Y. (2004). Distinct roles for PP1 and PP2A in the *Neurospora* circadian clock. *Genes Dev.* **18**, 255–260.

Yanovsky, M. J., and Kay, S. A. (2002). Molecular basis of seasonal time measurement in *Arabidopsis*. *Nature* **419**, 308–312.

Young, M. W., and Kay, S. A. (2001). Time zones: A comparative genetics of circadian clocks. *Nat. Rev. Genet.* **2**, 702–715.

3

Involvement of Homologous Recombination in Carcinogenesis

Ramune Reliene,* Alexander J. R. Bishop,§ and Robert H. Schiestl*,†,‡

*Department of Pathology, Geffen School of Medicine
Los Angeles, California 90024
†Department of Environmental Health, School of Public Health
Los Angeles, California 90024
‡Department of Radiation Oncology, Geffen School of Medicine
Los Angeles, California 90024
§Department of Cellular and Structural Biology, Children's Cancer Research
Institute, University of Texas Health Science Center San Antonio
San Antonio, Texas 78229

Advances in Genetics, Vol. 58
Copyright 2007, Elsevier Inc. All rights reserved.

0065-2660/07 $35.00
DOI: 10.1016/S0065-2660(06)58003-4

ABSTRACT

DNA alterations of every type are associated with the incidence of carcinogenesis, often on the genomic scale. Although homologous recombination (HR) is an important pathway of DNA repair, evidence is accumulating that deleterious genomic rearrangements can result from HR. It therefore follows that HR events may play a causative role in carcinogenesis. HR is elevated in response to carcinogens. HR may also be increased or decreased when its upstream regulation is perturbed or components of the HR machinery itself are not fully functional. This chapter summarizes research findings that demonstrate an association between HR and carcinogenesis. Increased or decreased frequencies of HR have been found in cancer cells and cancer-prone hereditary human disorders characterized by mutations in genes playing a role in HR, such as ATM, Tp53, BRCA, BLM, and WRN genes. Another evidence linking perturbations in HR and carcinogenesis is provided by studies showing that exposure to carcinogens results in an increased frequency of HR resulting in DNA deletions in yeast, human cells, or mice.   © 2007, Elsevier Inc.

I. INTRODUCTION

An increased risk of developing cancer has been associated with a number of factors. Such factors include inherited or acquired mutations in tumor suppressor genes, proto-oncogenes and/or DNA damage response and repair genes, long-term environmental or occupational exposure to carcinogens, infection by retroviruses, and aging. Genetic alterations have been associated with carcinogenesis under all of these circumstances.

The process of carcinogenesis can be explained by several accepted models, one-step, two-step, and multistep models. In a one-step model, a dominant gene mutation occurs in a proto-oncogene whereby its oncogenic potential is revealed allowing carcinogenesis (Todd and Wong, 1999). Alternatively, inactivation of a tumor suppressor gene in heterozygous state may also promote carcinogenesis in one-step (Macleod, 2000). Such inactivation of a functional counterpart, termed loss of heterozygosity (LOH), is frequently found in human tumors (Davies *et al.*, 1999; Lohmann, 1999; Mitelman *et al.*, 2003). In a two-step model, both copies of a functional tumor suppressor gene become inactivated (Knudson, 1993). In a multistep model, proto-oncogenes and tumor suppressor genes may be intact, but defects in DNA damage response and repair allow accumulation of multiple gene mutations, gross genomic rearrangements, and LOH that may lead to activation of proto-oncogenes and/or loss of tumor suppressor genes (Heppner and Miller, 1998; Loeb and Loeb, 2000).

Human tumors as well as cancer-prone hereditary disorders are characterized by a wide variety of genome rearrangements such as deletions, translocations, duplication, as well as LOH and aneuploidy (Knudson, 2001; Mitelman et al., 2003). Evidence is accumulating that homologous recombination (HR) takes part in the formation of such genetic instability, although, first of all, we would like to present a beneficial role of HR.

II. HR INVOLVEMENT IN DNA REPAIR

DNA double-strand breaks (DSBs) that occur in our genome under various circumstances (e.g., exposure to DNA damaging agents or topoisomerase II inhibitors, during antigen receptor gene rearrangements in lymphocytes or DNA replication) are repaired by HR and nonhomologous end-joining (NHEJ) pathways (Haber, 2000; Jackson, 2002; Pierce and Jasin, 2001; Thompson and Schild, 2001). HR is a high-fidelity pathway of DNA repair by virtue of its use of a homologous piece of DNA as its template for the repair reaction. In contrast, NHEJ is inherently error-prone, often resulting in deletions of sequences around the break point.

The HR reaction is now understood to be a multistep process involving several complexes of proteins recruited in a stepwise manner (Krogh and Symington, 2004). Relevant proteins include RAD51, a homologue of bacterial RecA, RAD51-like proteins, RAD52, RAD54, and BRCA2 (Henning and Sturzbecher, 2003; Thacker, 2005; Thompson and Schild, 2002). In eukaryotes, RAD51 is a central protein of HR, which interacts directly or indirectly with other recombination proteins (RAD52, RAD54, RAD51-like proteins, BRCA2, BLM, WRN) and DNA damage sensors (ATM, p53, FANC proteins, BRCA1, and BRCA2). In addition to RAD51, there are the RAD51-like proteins (XRCC2, XRCC3, RAD51L1, RAD51L2, RAD51L3), often called RAD51 paralogues, that is proteins derived from a single gene after its duplication in the genome, although the sequence identity with RAD51 is relatively low with the highest conservation being in the putative ATP-binding domain (Thacker, 2005).

HR-mediated repair of DNA DSBs is best understood in the yeast Saccharomyces cerevisiae (Krogh and Symington, 2004). A DSB is an initiating lesion, where DNA strand resection occurs allowing the formation of RAD51 filaments on exposed 3′ single-stranded DNA tails. Similar to bacterial RecA, RAD51 stretches out the single-stranded DNA, protects it, and facilitates its single-stranded invasion of a double-stranded DNA homologue. Subsequent steps include DNA synthesis primed from this single-stranded DNA end using undamaged complementary homologous sequence from the invaded double-stranded DNA, possibly strand migration along this template and finally an enzymatic resolution of resultant cross-stranded structure called a Holliday junction (Fig. 3.1). HR is also capable of repairing interstrand cross-links and DNA adducts in close proximity on opposite strands. In addition, HR resolves stalled replication forks that are routinely arrested by an even greater variety of stresses (template lesions, template-bound proteins, nucleotide depletion, temperature changes) (Hyrien, 2000; Kuzminov, 2001) that may produce DNA breakage.

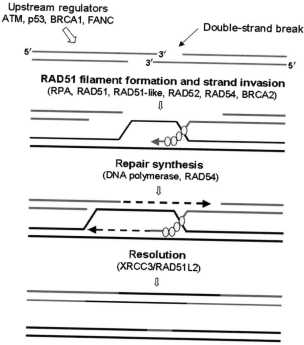

Figure 3.1. DNA DSB repair by HR. RAD51 forms filaments on single-stranded 3′ DNA after displacement of single-stranded DNA-binding protein RPA and catalyzes strand transfer between broken sequence and its homologue (strand invasion) to allow resynthesis of the damaged DNA fragment (Krogh and Symington, 2004). RAD52 and probably RAD51-like proteins assist displacement of RPA and annealing of the invaded strand to the undamaged complementary homologous strand. After repair synthesis, Holliday junctions are resolved possibly by XRCC3/RAD51L2 heterodimer. RAD54 and RAD51-like proteins seem to play a role before and after stand invasion to facilitate opening up of DNA and removal of proteins bound to single-stranded DNA. The role of HR regulators is less understood. Some of these proteins are involved in upstream steps of DNA damage response, while BLM and WRN appear to be involved in branch migration of Holliday junction, as discussed below. (See Color Insert.)

III. HR AND CARCINOGENESIS

HR provides a means of copying genetic information from undamaged template present on either a sister chromatid or homologous chromosome. By using this undamaged template, a high level of fidelity can be ensured. However, the genome contains many repetitive elements, and inappropriate template usage during HR repair may be deleterious, and may result in deletions, translocations, duplication, LOH, or aneuploidy (Lengauer et al., 1998; Zhou et al., 2001). Such inappropriate recombination mostly occurs when HR is disturbed by deficiency in HR machinery genes, DNA damage sensors and cell cycle regulators, or environmental agents causing DNA damage. Examples of these genes include BRCA1, BRCA2, ATM, Tp53, FANC, BLM, and WRN genes. Cells that carry such mutations display a high level of genetic instability and sensitivity to cross-linking agents and ionizing radiation (Thompson and Schild, 2002). Such mutations cause hereditary cancer-prone syndromes.

A. RAD51

RAD51 is a central protein in DNA repair by HR (Thacker, 2005). The importance of HR and the indispensable role of RAD51 and RAD51-like proteins is underscored by the observation that mice lacking Rad51, and thereby the capacity to perform HR, die during early embryonic development (Lim and Hasty, 1996; Tsuzuki et al., 1996). In comparison, mice deficient in the Rad51-like genes, Xrcc2 (Deans et al., 2000), Rad51L1 (Shu et al., 1999), Rad51L3 (Pittman and Schimenti, 2000) progress during development but no live offspring are found. Similarly, it was not possible to establish cell cultures from Rad51-deficient mouse embryos (Lim and Hasty, 1996), while mouse embryonic fibroblasts (MEFs) and ES cells derived from Xrcc2 knockout mice display a high frequency of chromosomal instability (Deans et al., 2000). Overexpression of RAD51 in cells can stimulate (Arnaudeau et al., 1999; Vispe et al., 1998) or reduce HR (Kim et al., 2001) and potentially lead to genome rearrangements (Richardson et al., 2004). Experimental data provide some evidence that alterations in RAD51 and RAD51 family genes may play a role in carcinogenesis. However, to date, any association between cancer in humans and polymorphisms, mutations or differential expression levels of RAD51 are weak (reviewed in Thacker, 2005). For example, the RAD51L3 variant, Glu 233Gly (Rodriguez-Lopez et al., 2004), and XRCC2 variant, Arg188His (Rafii et al., 2002), may marginally increase breast cancer; XRCC3 variants in some studies increased but in other studies had no effect on the risk of breast cancer, basal cell carcinoma, or malignant melanoma (Thacker, 2005). Decreased RAD51 expression was found in 30% of 179 breast cancer cases, mostly sporadic and some being associated with BRCA1 mutations (Yoshikawa et al., 2000). In contrast, increased RAD51 expression correlated with sporadic invasive ductal breast cancer (Maacke *et al.*, 2000a) or pancreatic adenocarcinoma (Maacke *et al.*, 2000b). Alterations in other *RAD51* family genes are only suggestive of increased cancer risk. Much better established are alterations in other genes involved in HR, for example *BRCA1* or *BRCA2*.

B. BRCA

Inherited mutations in *BRCA1* and *BRCA2* genes predispose to breast, ovarian, and other cancers (Murphy *et al.*, 2002; Nathanson *et al.*, 2001; Scully and Livingston, 2000). One defective copy is enough to cause a predisposition to cancer, although the loss of the second allele is consistently found in tumors from the patients carrying *BRCA* germ line mutations. Cells deficient in *BRCA* genes exhibit high level of chromosomal breakage, deletions, centrosome defects, and aneuploidy (Kraakman-van der Zwet *et al.*, 2002; Tutt *et al.*, 1999; Venkitaraman, 2001; Xu *et al.*, 1999). *BRCA*-deficient cells are specifically deficient in HR, while NHEJ remains intact (Moynahan *et al.*, 2001; Snouwaert *et al.*, 1999). BRCA proteins are essential in HR, although the exact role in this process is not known. BRCA2 interacts with RAD51 via a series of BRC repeats (Wong *et al.*, 1997) and thereby might promote targeting and oligomerization of RAD51 on single-stranded DNA during an HR event (Shivji and Venkitaraman, 2004).

BRCA1 is known to interact with many other proteins (Wang *et al.*, 2000b) and as such may have a role early in dictating or coordinating an HR repair response. For example, BRCA1 has been demonstrated to be involved in signaling cell cycle delay on DNA damage (Xu *et al.*, 1999) and, as such, may provide an opportunity for HR repair to be enacted.

C. ATM

ATM (ataxia-telangiectasia-mutated) protein has multiple functions and is involved in HR as an upstream regulator. Bi-allelic mutations in the *ATM* gene cause a human disorder ataxia-telangiectasia (AT) characterized by a variety of clinical manifestations, such as early onset neurological degeneration, ocular telangiectasias (dilation of blood capillaries), immunodeficiency, growth retardation, sterility, radiosensitivity, and extremely high predisposition to cancer particularly in lymphoid tissues (Becker-Catania and Gatti, 2001; Lavin and Shiloh, 1997; Meyn, 1999). ATM-deficient cells display oxidative stress and various forms of genetic instability such as chromosome breaks, acentric fragments, dicentric chromosomes, aneuploidy, and translocations (Cohen and Levy, 1989; Kojis *et al.*, 1991; Reichenbach *et al.*, 1999, 2002; Stumm *et al.*, 2001). Hyperreactivity in HR has been observed in *Atm* knockout mice (Bishop *et al.*, 2000a, 2003; Reliene *et al.*, 2004). Such hyperreactivity leads to intrachromosomal HR between repeated sequences and results in DNA deletions. The ATM protein, a phosphotidylinositol 3-like kinase, is considered a primary activator in cellular response to DNA DSBs after initial DSB detection by the MRN complex (Carson *et al.*, 2003; Uziel *et al.*, 2003). On DNA damage, ATM phosphorylates its substrates, for example p53 and Chk1/2, leading to cell cycle arrest allowing time for repair. ATM also phosphorylates proteins involved in the HR pathway, such as BRCA1, c-Abl, which further directly or indirectly interact with RAD51 and RAD52. Thus, ATM coordinates HR repair via multiple interactions with other relevant members of the DNA damage response pathway. In addition, ATM has been implicated in response to oxidative stress because a lack of the functional ATM protein leads to oxidative stress and oxidative damage in various organs, cells, and macromolecules (Barlow *et al.*, 1999; Kamsler *et al.*, 2001; Reichenbach *et al.*, 2002). Oxidative damage to DNA is associated with hyperrecombination, as has been shown in *Atm* knockout mice (Reliene *et al.*, 2004). Interestingly, addition of antioxidant *N*-acetyl cystein to the mouse diet prevented oxidative DNA damage and hyperrecombination (Reliene *et al.*, 2004). Thus, ATM protein regulates HR through its dual and possibly overlapping function such as a DNA damage response and oxidative stress response.

D. Tp53

Li–Fraumeni syndrome is a dominantly inherited disorder caused by mutations in the tumor suppressor gene *Tp53* (Li and Fraumeni, 1969; Strong *et al.*, 1992). Tumorigenesis in Li–Fraumeni patients is associated with the loss of the remaining functional allele and nearly 100% of patients develop cancer. A variety of tumors, such as breast cancer, lymphoma, lung carcinoma, brain cancer, adenocortical carcinoma usually appear in children and young adults. Cells deficient in the protein encoded by *Tp53*, p53, are not capable to arrest cell growth after exposure to DNA-damaging agents and display high frequency of spontaneous chromosomal rearrangements and elevated rates of spontaneous as well as induced HR (Aubrecht *et al.*, 1999; Bertrand *et al.*, 1997; Gebow *et al.*, 2000; Levine, 1997; Livingstone *et al.*, 1992; Mekeel *et al.*, 1997; Yin *et al.*, 1992). It has been postulated that p53 is involved in suppression of excessive HR via cell cycle regulation in the common ATM-p53 DNA damage response pathway, though, at least spontaneously, its most potent cell cycle effector, p21, plays no role regulating HR frequency (Bishop *et al.*, 2006). Besides that, p53 seems to have a more direct role in HR via interactions with proteins implicated in the HR machinery, such as BRCA1, BRCA2, BLM, RAD52, RAD51, and RPA (Buchhop *et al.*, 1997; Marmorstein *et al.*, 1998; Sturzbecher *et al.*, 1996; Wang *et al.*, 2001; Zhang *et al.*, 1998). In fact, it has been demonstrated that p53 can interact with RAD51 possibly interfering with RAD51 homodimerization or RAD51–RPA interaction (Buchhop *et al.*, 1997; Sturzbecher *et al.*, 1996).

In either case, it is interesting to note that from our studies, the absence of p53 results in an increased level of HR, particularly in the early part of development (Bishop *et al.*, 2003). In the same study, we observed that the absence of Gadd45a also resulted in an increased level of HR with a surprisingly similar profile to that of p53. This is particularly interesting considering the interactions of Gadd45a with BRCA1 (for a review see Bishop *et al.*, 2003) and may suggest that indeed the p53 contribution to the regulation of HR is through protein interactions.

E. BLM

Bloom's syndrome (BS) is a rare autosomal recessive disorder due to inherited mutations in the *BLM* gene. The BS patients are characterized by stunted growth, sensitivity to sunlight, immunodeficiency, fertility defects, and greatly elevated incidence of various cancers (German, 1993). BS patients develop cancers that are common in the general population, such as leukemia, lymphoma, various types of carcinomas, and those that are rare in the general populations, such as Wilms tumor and osteosarcoma. Cells from BS patients exhibit elevated sister chromatid exchange events indicating hyperrecombination (Ellis *et al.*, 1995; Hickson *et al.*, 2001; Wang *et al.*, 2000a). Cultured BS cells display greatly increased levels of chromatid breaks, gaps, and rearranged chromosomes. The *BLM* gene encodes a member of the RecQ family of helicases (Ellis *et al.*, 1995). The action of this helicase appears to be required for proficient HR, particularly in repairing stalled replication forks. One study has suggested that BLM helicase is involved in promoting branch migration of Holliday junctions, a recombination intermediate, facilitating their elimination as required for the completion of HR (Karow *et al.*, 2000). Association between BLM and HR is evidenced by interaction between BLM and RAD51 *in vitro* and in coimmunoprecipitates from nuclear extracts (Wu *et al.*, 2001).

F. WRN

A rare autosomal recessive disorder Werner syndrome (WS) is caused by inherited mutations in the *WRN* gene that encodes a RecQ helicase homologous to BLM helicase (Shen and Loeb, 2001). Like BS patients, WS patients are short in statue and prone to carcinogenesis at various sites. Unlike BS, WS is an adult progeria syndrome characterized by an early onset of age-related traits, including osteoporosis, cataracts, atherosclerosis, diabetes mellitus, and loss of skin elasticity and hair. WRN-deficient cells show greatly increased genetic instability, including deletions, translocations, and inversions, although, unlike *BLM*-deficient cells, have normal frequency of sister chromatid exchange (Imamura *et al.*, 2002; Oshima *et al.*, 2002; Shen and Loeb, 2001). Research findings suggest that WRN helicase plays a role in resolving recombination intermediates (Constantinou et al., 2000). The absence of WRN results in an increased frequency of DNA deletions as a consequence of inappropriate HR in mouse models (Lebel, 2002).

G. FANC

Autosomal recessive mutations in FANC genes cause a human disorder, Fanconi's anemia (FA), characterized by bone marrow failure, specific skeletal defects, and a high incidence of cancers, predominantly acute myelogenous leukemia and squamous cell carcinoma (Joenje and Patel, 2001). Cells of FA patients exhibit chromosomal instability and enhanced sensitivity to bifunctional alkylating agents that cross-link DNA (Carreau et al., 1999). FANC-deficient cells display mild defects in HR. Contradictions in the literature report either impaired HR (Digweed et al., 2002) or elevated HR (Donahue et al., 2003). There are at least eight FA complementation groups (A, B, C, D1, D2, E, F, and G) most of whose genes have been cloned (Joenje and Patel, 2001). The FANCA, FANCC, FANCE, FANCF, and FANCG proteins assemble in a multisubunit complex functioning as a monoubiquitine ligase, required for monoubiquitination of the FANCD2 protein (Garcia-Higuera et al., 2001). When a cell is exposed to a DNA-damaging agent, this monoubiquitinated FANCD2 is targeted to nuclear foci, where it has been demonstrated to interact with BRCA1 and RAD51 (Taniguchi et al., 2002). It has further been demonstrated that this FANCD2 monoubiquitination is critical for normal levels of HR (Nakanishi et al., 2005). More recently, the FA pathway is suggested to be involved in the stabilization of stalled or broken replication forks, which facilitates HR-mediated restoration of replication (Thompson et al., 2005).

IV. OTHER DISEASES ATTRIBUTABLE TO HR

Repetitive DNA sequences comprise about 25% of the mammalian genome (Schmid et al., 1989). This provides many opportunities for inappropriate HR between intrachromosomal sites, which, as a result, may result in deletions, duplications, or gene conversions (Calabretta et al., 1982). Several diseases have been attributed to deletions of large DNA fragments as a consequence of HR between repeated sequences (Ji et al., 2000; Lupski, 1998; Mazzarella and Schlessinger, 1998; Purandare and Patel, 1997; Reliene and Schiestl, 2003). These include X-linked ichthyosis (1.9 Mb deleted by flanking S232 elements) (Ballabio et al., 1990; Yen et al., 1990), Prader–Willi syndrome (Ledbetter et al., 1981), DiGeorge syndrome (de la Chapelle et al., 1981), and hypercholesterolemia (Lehrman et al., 1985). A tandem duplication mediated by Alu recombination within the *ALL-1* gene is considered the causative event that resulted in an acute myeloid leukemia (Schichman *et al.*, 1994). Charcot–Marie–Tooth syndrome 1A and hereditary neuropathy with liability to pressure palsies are often caused by a chromosome duplication or deletion, respectively, of a 1.5-Mb region flanked by 24 kb tandemly repeated sequences (Chance *et al.*, 1993; Nelis *et al.*, 1996; Pentao *et al.*, 1992; Wise *et al.*, 1993). An HR-mediated inversion disrupting the factor VIII gene, *F8*, is a cause in about 45% of severe hemophilia A cases (Naylor *et al.*, 1993, 1996).

V. CARCINOGEN-INDUCED HR IN EXPERIMENTAL SYSTEMS

In this section, we present results obtained in our laboratory on carcinogen-induced elevated frequency of HR, which provides further evidence on the association of HR and carcinogenesis. We have constructed and/or used HR assays in yeast, human cells, and mice to score for homology-mediated deletion events between duplicated DNA sequences. An HR substrate was a duplicated region of the *his3* allele in yeast *S. cerevisiae* (Schiestl *et al.*, 1988), a duplication of exons 2 and 3 of the *hprt* gene in a human lymphoblastoma cell line (Aubrecht *et al.*, 1995), and a duplication of a 70-kb fragment spanning exons 6–18 of the *pink-eyed dilution* (*p*) gene in mice (Schiestl *et al.*, 1994). A number of carcinogens induced an elevated frequency of DNA deletions in the yeast, human cells, and/or mouse assay by an HR-mediated mechanism (Table 3.1; Fig. 3.2). These include ionizing and UV radiation, isolated chemical carcinogens as well as compound mixture carcinogens such as cigarette smoke or diesel exhaust particles (DEP). Some of these carcinogens specifically induce recombination events, but do not cause point mutations, as tested by a classical mutagenicity assay (Ames) in bacteria *S. thyphimurium*. Benzene, arsenate, polychlorinated biphenyl, TCDD, urethane, and acetamide are just a few examples of carcinogens that have been tested positive in the above-mentioned HR assays and negative in the Ames assay. The tested carcinogens cause DNA damage by various mechanisms. Some of them cross-link, break down, or alkylate DNA, others generate free radicals and thereby induce oxidative DNA lesions. Interestingly, diverse Ames assay-negative carcinogens, such as cadmium, aniline, chloroform, carbon tetrachloride, benzene, urethane, and thiourea, induce HR by a free radical mechanism, as has been shown in yeast (Brennan and Schiestl, 1996, 1997a, 1998a,b).

Other similar assays also exist measuring other forms of HR such as the *Aprt*$^{+/-}$ mouse for measuring gross LOH events (Wijnhoven *et al.*, 2001) and the FYDR mouse for measuring gene conversion events (Kovalchuk *et al.*, 2004).

Studies on human and experimental carcinogenesis have revealed that cell proliferation is an important factor for tumor formation. Our studies on

Table 3.1. Inducibility of Homology-Mediated Deletions by Carcinogens in the Yeast, Human Cells, and Mouse Assays[a]

Carcinogen	DEL assay Yeast	DEL assay Human cells	DEL assay Mouse
2,4-Diaminotoluene (Brennan and Schiestl, 1997b)	+		
3-Amino-1,2,4-triazole (Schiestl et al., 1989a)	+		
4-NQO (Schiestl et al., 1989a)	+		
Acrylonitrile (Carls and Schiestl, 1994)	+		
Aflatoxin B1 (Schiestl et al., 1989a)	+		
Aniline (Brennan and Schiestl, 1998b; Schiestl et al., 1989a)	+		
Aroclor 1260 (PCB) (Schiestl et al., 1997a)			+
Aroclor 1221 (PCB) (Schiestl et al., 1997a)	+	+	+
Arsenate (Schiestl et al., 1997b)			+
Auramine O (Brennan and Schiestl, 1998b; Galli and Schiestl, 1996)	+		+
Benzene (Aubrecht et al., 1995; Brennan and Schiestl, 1998b; Schiestl et al., 1989a, 1997b)	+	+	+
Benzo(a)pyrene (Bishop et al., 2000; Schiestl et al., 1997b)			+
Cadmium chloride (Brennan and Schiestl, 1998b; Schiestl et al., 1989a)	+		
Carbon tetrachloride (Schiestl et al., 1989a)	+		
Cyclophosphamide (Carls and Schiestl, 1994)	+		
DDE (Schiestl et al., 1989a)	+		
Dimethylhydrazine (Schiestl et al., 1989a)	+		
EMS (Brennan and Schiestl, 1998b; Galli and Schiestl, 1999, p. 131; Schiestl et al., 1989a, 1997b)	+		+
ENU (Schiestl et al., 1997b)			+
Epichlorohydrin (Schiestl et al., 1989a)	+		
Ethionine (Schiestl et al., 1989a)	+		
Ethylene dibromide (Schiestl et al., 1989a)	+		
Ethylenethiourea (Schiestl et al., 1989a)	+		
Formaldehyde (Schiestl et al., 1989a)	+		

(Continues)

Table 3.1. (Continued)

Carcinogen	DEL assay			Carcinogen	DEL assay		
	Yeast	Human cells	Mouse		Yeast	Human cells	Mouse
Hexamethyl phosphoramide (Carls and Schiestl, 1994)	+			o-Toluidine (Brennan and Schiestl, 1999)	+		
Ionizing radiation (Aubrecht et al., 1995; Bishop et al., 2000b; Schiestl et al., 1989a, 1994)	+	+	+	Tobacco smoke (Jalili et al., 1998)			+
Methyl eugenol (Schiestl et al., 1989b)	+			TCDD (Schiestl et al., 1997a)			+
Methylene chloride (Schiestl et al., 1989a)	+			Thioacetamide (Schiestl et al., 1989a)	+		
MMS (Aubrecht et al., 1995; Brennan and Schiestl, 1998b; Galli and Schiestl, 1999; Schiestl et al., 1989a, 1997b)	+	+	+	Thiourea (Aubrecht et al., 1995; Brennan and Schiestl, 1998b; Schiestl et al., 1989a)	+	+	
Nitrogen mustard (Schiestl et al., 1989a)	+			Urethane (Brennan and Schiestl, 1998b; Schiestl et al., 1989a)	+		
o-Anisidine (Brennan and Schiestl, 1999)	+			UV irradiation (Aubrecht et al., 1995; Schiestl et al., 1989a)	+	+	
				DEP (Reliene et al., 2005)			+

[a] Chemical abbreviations: DDE, 2,2-bis(4chlorophenyl)-1,1dichloroethylene; 4-NQO, 4-nitroquinoline-N-oxide; ENU, 1-ethyl-1-nitrosourea; MMS, methyl methanesulfonate; EMS, ethyl methanesulfonate; TPA, 12-O-tetradecanoylphorbol-13-acetate; PCB, polychlorinated biphenyl; TCDD, 2,3,7,8-tetra-chlorodibenzo-p-dioxin; DEP, diesel exhaust particles.

Data for carcinogenicity can be found in the US Environmental Protection Agency Integrated Risk Information System (http://www.epa.gov/iris/subst) or in the Carcinogenesis Potency Database (http://potency.berkeley.edu/chemicalsummary.html).

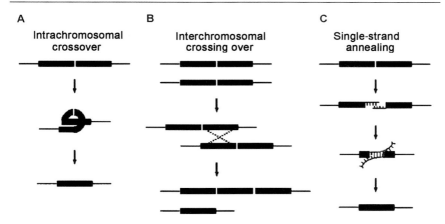

Figure 3.2. Mechanisms of homology-mediated deletions between duplicated DNA sequences. (A) Intrachromosomal crossover. Crossover events occur following alignment of homologous sequences, strand invasion, and strand exchange. (B) Unequal crossing over between homologous chromosomes or sister chromatids. (C) Single-strand annealing. A single-stranded exonuclease degrades 5′ ends of a double-stranded DNA at a DSB until complementary homologous sequences on the remaining 3′ single-stranded DNA tails are exposed and can anneal. Single-strand annealing is an alternative pathway of homology-mediated repair of DSBs, which leads to a loss of DNA fragment due to degradation of 5′ end. Adapted from Reliene and Schiestl (2003).

carcinogen-induced HR summarized in Table 3.1 demonstrated that proliferating cells are highly prone to accumulation of DNA deletions generated by HR. This has been concluded because proliferating cells underwent HR-mediated deletions following exposure to carcinogens causing various forms of DNA damage, being it single-strand breaks (SSBs), DSBs, UV lesions, or alkylation, while cell cycle-arrested cells showed elevated frequency of deletions only after exposure to DNA DSB inducing carcinogen (ionizing irradiation) (Galli and Schiestl, 1999). However, when arrested cells were allowed to progress through cell cycle, an elevated frequency of deletions was observed irrespective of lesion type. Apparently, the hyperactivity in HR in dividing cells results from DNA replication during which template lesions are converted into DSBs, which then lead to HR and cause DNA rearrangements.

VI. SUMMARY

In this chapter, we reviewed research findings evidencing that HR is associated with carcinogenesis. Under normal conditions, HR is a highly regulated process designed to repair DNA damages, particularly DSBs. As evidenced here, there are a number of genes involved in controlling and executing the HR process. Some of these genes encode constituents of the HR machinery such as RAD51, RAD52, and RAD54; others are involved in the regulation of HR, exemplified by ATM, p53, and BRCA1. This latter class of proteins are not only involved in HR, but play a role in coordinating various cellular processes including cell cycle regulation and DNA damage response, and generally, in the maintenance of genome integrity and tumor suppression. As such, these signaling pathways highlight the interactive nature of damage response and how HR is a controlled event on a system-wide level. The deregulation of one component at almost any level can have profound effects on HR, resulting in either its impairment or hyperactivity. The association between HR and carcinogenesis has been demonstrated by findings that increased or decreased frequencies of HR are found in cancer cells and cancer-prone hereditary human disorders characterized by mutations in *ATM, Tp53, BRCA, BLM, WRN,* and *FANC* genes. This evidence is compounded by the finding that carcinogen exposure increases HR thus facilitating deleterious genomic rearrangements. Together it is now understood that HR provides a powerful and necessary repair mechanism when correctly utilized. Inappropriate HR, in contrast, can produce deleterious genomic rearrangements that facilitate carcinogenesis.

Acknowledgments

Supported by grants from the National Institute of Environmental Health Sciences, NIH, RO1 grant No. ES09519 (to RHS), postdoctoral research fellowship of the University of California Toxic Substances Research and Teaching Program, the Lymphoma Research Foundation Elizabeth Banks Jacobs & Byron Wade Strunk Memorial Fellowship (both to RR), and NIH/NIEHS award (K22 ES 012264) to AJRB.

References

Arnaudeau, C., Helleday, T., and Jenssen, D. (1999). The RAD51 protein supports homologous recombination by an exchange mechanism in mammalian cells. *J. Mol. Biol.* **289,** 1231–1238.

Aubrecht, J., Rugo, R., and Schiestl, R. H. (1995). Carcinogens induce intrachromosomal recombination in human cells. *Carcinogenesis* **16,** 2841–2846.

Aubrecht, J., Secretan, M. B., Bishop, A. J., and Schiestl, R. H. (1999). Involvement of p53 in X-ray induced intrachromosomal recombination in mice. *Carcinogenesis* **20,** 2229–2236.

Ballabio, A., Bardoni, B., Guioli, S., Basler, E., and Camerino, G. (1990). Two families of low-copy-number repeats are interspersed on Xp22.3: Implications for the high frequency of deletions in this region. *Genomics* **8,** 263–270.

Barlow, C., Dennery, P. A., Shigenaga, M. K., Smith, M. A., Morrow, J. D., Roberts, L. J., II, Wynshaw-Boris, A., and Levine, R. L. (1999). Loss of the ataxia-telangiectasia gene product causes oxidative damage in target organs. *Proc. Natl. Acad. Sci. USA* **96,** 9915–9919.

Becker-Catania, S. G., and Gatti, R. A. (2001). Ataxia-telangiectasia. *Adv. Exp. Med. Biol.* **495,** 191–198.

Bertrand, P., Rouillard, D., Boulet, A., Levalois, C., Soussi, T., and Lopez, B. S. (1997). Increase of spontaneous intrachromosomal homologous recombination in mammalian cells expressing a mutant p53 protein. *Oncogene* **14**, 1117–1122.

Bishop, A. J. R., Barlow, C., Wynshaw-Boris, A. J., and Schiestl, R. H. (2000a). Atm deficiency causes an increased frequency of intrachromosomal homologous recombination in mice. *Cancer Res.* **60**, 395–399.

Bishop, A. J. R., Kosaras, B., Sidman, R. L., and Schiestl, R. H. (2000b). Benzo(a)pyrene and X-rays induce reversions of the pink-eye unstable mutation in the retinal pigment epithelium of mice. *Mutat. Res.* **457**, 31–40.

Bishop, A. J., Hollander, M. C., Kosaras, B., Sidman, R. L., Fornace, A. J., Jr., and Schiestl, R. H. (2003). Atm-, p53-, and Gadd45a-deficient mice show an increased frequency of homologous recombination at different stages during development. *Cancer Res.* **63**, 5335–5343.

Bishop, A. J., Kosaras, B., Hollander, M. C., Fornace, A., Jr., Sidman, R. L., and Schiestl, R. H. (2006). p21 controls patterning but not homologous recombination in RPE development. *DNA Repair* **5**, 111–120.

Brennan, R. J., and Schiestl, R. H. (1996). Cadmium is an inducer of oxidative stress in yeast. *Mutat. Res.* **356**, 171–178.

Brennan, R. J., and Schiestl, R. H. (1997a). Aniline and its metabolites generate free radicals in yeast. *Mutagenesis* **12**, 215–220.

Brennan, R. J., and Schiestl, R. H. (1997b). Diaminotoluenes induce intrachromosomal recombination and free radicals in *Saccharomyces cerevisiae*. *Mutat. Res.* **381**, 251–258.

Brennan, R. J., and Schiestl, R. H. (1998a). Chloroform and carbon tetrachloride induce intrachromosomal recombination and oxidative free radicals in *Saccharomyces cerevisiae*. *Mutat. Res.* **397**, 271–278.

Brennan, R. J., and Schiestl, R. H. (1998b). Free radicals generated in yeast by the *Salmonella* test-negative carcinogens benzene, urethane, thiourea and auramine O. *Mutat. Res.* **403**, 65–73.

Brennan, R. J., and Schiestl, R. H. (1999). The aromatic amine carcinogens o-toluidine and o-anisidine induce free radicals and intrachromosomal recombination in *Saccharomyces cerevisiae*. *Mutat. Res.* **430**, 37–45.

Buchhop, S., Gibson, M. K., Wang, X. W., Wagner, P., Sturzbecher, H. W., and Harris, C. C. (1997). Interaction of p53 with the human Rad51 protein. *Nucleic Acids Res.* **25**, 3868–3874.

Calabretta, B., Robberson, D. L., Barrera-Saldana, H. A., Lambrou, T. P., and Saunders, G. F. (1982). Genome instability in a region of human DNA enriched in Alu repeat sequences. *Nature* **296**, 219–225.

Carls, N., and Schiestl, R. H. (1994). Evaluation of the yeast DEL assay with 10 compounds selected by the International Program on Chemical Safety for the evaluation of short-term tests for carcinogens. *Mutat. Res.* **320**, 293–303.

Carreau, M., Alon, N., Bosnoyan-Collins, L., Joenje, H., and Buchwald, M. (1999). Drug sensitivity spectra in Fanconi anemia lymphoblastoid cell lines of defined complementation groups. *Mutat. Res.* **435**, 103–109.

Carson, C. T., Schwartz, R. A., Stracker, T. H., Lilley, C. E., Lee, D. V., and Weitzman, M. D. (2003). The Mre11 complex is required for ATM activation and the G2/M checkpoint. *EMBO J.* **22**, 6610–6620.

Chance, P. F., Alderson, M. K., Leppig, K. A., Lensch, M. W., Matsunami, N., Smith, B., Swanson, P. D., Odelberg, S. J., Disteche, C. M., and Bird, T. D. (1993). DNA deletion associated with hereditary neuropathy with liability to pressure palsies. *Cell* **72**, 143–151.

Cohen, M. M., and Levy, H. P. (1989). Chromosome instability syndromes. *Adv. Hum. Genet.* **18**, 43–149, 365–371.

Constantinou, A., Tarsounas, M., Karow, J. K., Brosh, R. M., Bohr, V. A., Hickson, I. D., and West, S. C. (2000). Werner's syndrome protein (WRN) migrates Holliday junctions and co-localizes with RPA upon replication arrest. *EMBO Rep.* **1,** 80–84.

Davies, R., Moore, A., Schedl, A., Bratt, E., Miyahawa, K., Ladomery, M., Miles, C., Menke, A., van Heyningen, V., and Hastie, N. (1999). Multiple roles for the Wilms' tumor suppressor, WT1. *Cancer Res.* **59,** 1747s–1750s; discussion 1751s.

Deans, B., Griffin, C. S., Maconochie, M., and Thacker, J. (2000). Xrcc2 is required for genetic stability, embryonic neurogenesis and viability in mice. *EMBO J.* **19,** 6675–6685.

de la Chapelle, A., Herva, R., Koivisto, M., and Aula, P. (1981). A deletion in chromosome 22 can cause DiGeorge syndrome. *Hum. Genet.* **57,** 253–256.

Digweed, M., Rothe, S., Demuth, I., Scholz, R., Schindler, D., Stumm, M., Grompe, M., Jordan, A., and Sperling, K. (2002). Attenuation of the formation of DNA-repair foci containing RAD51 in Fanconi anaemia. *Carcinogenesis* **23,** 1121–1126.

Donahue, S. L., Lundberg, R., Saplis, R., and Campbell, C. (2003). Deficient regulation of DNA double-strand break repair in Fanconi anemia fibroblasts. *J. Biol. Chem.* **278,** 29487–29495.

Ellis, N. A., Groden, J., Ye, T. Z., Straughen, J., Lennon, D. J., Ciocci, S., Proytcheva, M., and German, J. (1995). The Bloom's syndrome gene product is homologous to RecQ helicases. *Cell* **83,** 655–666.

Galli, A., and Schiestl, R. H. (1996). Effects of Salmonella assay negative and positive carcinogens on intrachromosomal recombination in G1-arrested yeast cells. *Mutat. Res.* **370,** 209–221.

Galli, A., and Schiestl, R. H. (1999). Cell division transforms mutagenic lesions into deletion-recombinagenic lesions in yeast cells. *Mutat. Res.* **429,** 13–26.

Garcia-Higuera, I., Taniguchi, T., Ganesan, S., Meyn, M. S., Timmers, C., Hejna, J., Grompe, M., and D'Andrea, A. D. (2001). Interaction of the Fanconi anemia proteins and BRCA1 in a common pathway. *Mol. Cell* **7,** 249–262.

Gebow, D., Miselis, N., and Liber, H. L. (2000). Homologous and nonhomologous recombination resulting in deletion: Effects of p53 status, microhomology, and repetitive DNA length and orientation. *Mol. Cell. Biol.* **20,** 4028–4035.

German, J. (1993). Bloom syndrome: A mendelian prototype of somatic mutational disease. *Medicine (Baltimore)* **72,** 393–406.

Haber, J. E. (2000). Partners and pathwaysrepairing a double-strand break. *Trends Genet.* **16,** 259–264.

Henning, W., and Sturzbecher, H. W. (2003). Homologous recombination and cell cycle checkpoints: Rad51 in tumour progression and therapy resistance. *Toxicology* **193,** 91–109.

Heppner, G. H., and Miller, F. R. (1998). The cellular basis of tumor progression. *Int. Rev. Cytol.* **177,** 1–56.

Hickson, I. D., Davies, S. L., Li, J. L., Levitt, N. C., Mohaghegh, P., North, P. S., and Wu, L. (2001). Role of the Bloom's syndrome helicase in maintenance of genome stability. *Biochem. Soc. Trans.* **29,** 201–204.

Hyrien, O. (2000). Mechanisms and consequences of replication fork arrest. *Biochimie* **82,** 5–17.

Imamura, O., Fujita, K., Itoh, C., Takeda, S., Furuichi, Y., and Matsumoto, T. (2002). Werner and Bloom helicases are involved in DNA repair in a complementary fashion. *Oncogene* **21,** 954–963.

Jackson, S. P. (2002). Sensing and repairing DNA double-strand breaks. *Carcinogenesis* **23,** 687–696.

Jalili, T., Murthy, G. G., and Schiestl, R. H. (1998). Cigarette smoke induces DNA deletions in the mouse embryo. *Cancer Res.* **58,** 2633–2638.

Ji, Y., Eichler, E. E., Schwartz, S., and Nicholls, R. D. (2000). Structure of chromosomal duplicons and their role in mediating human genomic disorders. *Genome Res.* **10,** 597–610.

Joenje, H., and Patel, K. J. (2001). The emerging genetic and molecular basis of Fanconi anaemia. *Nat. Rev. Genet.* **2,** 446–457.

Kamsler, A., Daily, D., Hochman, A., Stern, N., Shiloh, Y., Rotman, G., and Barzilai, A. (2001). Increased oxidative stress in ataxia telangiectasia evidenced by alterations in redox state of brains from Atm-deficient mice. *Cancer Res.* **61**, 1849–1854.

Karow, J. K., Constantinou, A., Li, J. L., West, S. C., and Hickson, I. D. (2000). The Bloom's syndrome gene product promotes branch migration of holliday junctions. *Proc. Natl. Acad. Sci. USA* **97**, 6504–6508.

Kim, P. M., Allen, C., Wagener, B. M., Shen, Z., and Nickoloff, J. A. (2001). Overexpression of human RAD51 and RAD52 reduces double-strand break-induced homologous recombination in mammalian cells. *Nucleic Acids Res.* **29**, 4352–4360.

Knudson, A. G. (1993). Antioncogenes and human cancer. *Proc. Natl. Acad. Sci. USA* **90**, 10914–10921.

Knudson, A. G. (2001). Two genetic hits (more or less) to cancer. *Nat. Rev. Cancer* **1**, 157–162.

Kojis, T. L., Gatti, R. A., and Sparkes, R. S. (1991). The cytogenetics of ataxia telangiectasia. *Cancer Genet. Cytogenet.* **56**, 143–156.

Kovalchuk, O., Hendricks, C. A., Cassie, S., Engelward, A. J., and Engelward, B. P. (2004). *In vivo* recombination after chronic damage exposure falls to below spontaneous levels in "recombomice." *Mol. Cancer Res.* **2**, 567–573.

Kraakman-van der Zwet, M., Overkamp, W. J., van Lange, R. E., Essers, J., van Duijn-Goedhart, A., Wiggers, I., Swaminathan, S., van Buul, P. P., Errami, A., Tan, R. T., Jaspers, N. G., Sharan, S. K., *et al.* (2002). Brca2 (XRCC11) deficiency results in radioresistant DNA synthesis and a higher frequency of spontaneous deletions. *Mol. Cell. Biol.* **22**, 669–679.

Krogh, B. O., and Symington, L. S. (2004). Recombination proteins in yeast. *Annu. Rev. Genet.* **38**, 233–271.

Kuzminov, A. (2001). DNA replication meets genetic exchange: Chromosomal damage and its repair by homologous recombination. *Proc. Natl. Acad. Sci. USA* **98**, 8461–8468.

Lavin, M. F., and Shiloh, Y. (1997). The genetic defect in ataxia-telangiectasia. *Annu. Rev. Immunol.* **15**, 177–202.

Lebel, M. (2002). Increased frequency of DNA deletions in pink-eyed unstable mice carrying a mutation in the Werner syndrome gene homologue. *Carcinogenesis* **23**, 213–216.

Ledbetter, D. H., Riccardi, V. M., Airhart, S. D., Strobel, R. J., Keenan, B. S., and Crawford, J. D. (1981). Deletions of chromosome 15 as a cause of the Prader-Willi syndrome. *N. Engl. J. Med.* **304**, 325–329.

Lehrman, M. A., Schneider, W. J., Sudhof, T. C., Brown, M. S., Goldstein, J. L., and Russell, D. W. (1985). Mutation in LDL receptor: Alu-Alu recombination deletes exons encoding transmembrane and cytoplasmic domains. *Science* **227**, 140–146.

Lengauer, C., Kinzler, K. W., and Vogelstein, B. (1998). Genetic instabilities in human cancers. *Nature* **396**, 643–649.

Levine, A. J. (1997). p53, the cellular gatekeeper for growth and division. *Cell* **88**, 323–331.

Li, F. P., and Fraumeni, J. F., Jr. (1969). Soft-tissue sarcomas, breast cancer, and other neoplasms. A familial syndrome? *Ann. Intern. Med.* **71**, 747–752.

Lim, D. S., and Hasty, P. (1996). A mutation in mouse rad51 results in an early embryonic lethal that is suppressed by a mutation in p53. *Mol. Cell. Biol.* **16**, 7133–7143.

Livingstone, L. R., White, A., Sprouse, J., Livanos, E., Jacks, T., and Tlsty, T. D. (1992). Altered cell cycle arrest and gene amplification potential accompany loss of wild-type p53. *Cell* **70**, 923–935.

Loeb, K. R., and Loeb, L. A. (2000). Significance of multiple mutations in cancer. *Carcinogenesis* **21**, 379–385.

Lohmann, D. R. (1999). RB1 gene mutations in retinoblastoma. *Hum. Mutat.* **14**, 283–288.

Lupski, J. R. (1998). Genomic disorders: Structural features of the genome can lead to DNA rearrangements and human disease traits. *Trends Genet.* **14**, 417–422.

Maacke, H., Opitz, S., Jost, K., Hamdorf, W., Henning, W., Kruger, S., Feller, A. C., Lopens, A., Diedrich, K., Schwinger, E., *et al.* (2000a). Over-expression of wild-type Rad51 correlates with histological grading of invasive ductal breast cancer. *Int. J. Cancer* **88**, 907–913.

Maacke, H., Jost, K., Opitz, S., Miska, S., Yuan, Y., Hasselbach, L., Luttges, J., Kalthoff, H., and Sturzbecher, H. W. (2000b). DNA repair and recombination factor Rad51 is over-expressed in human pancreatic adenocarcinoma. *Oncogene* **19**, 2791–2795.

Macleod, K. (2000). Tumor suppressor genes. *Curr. Opin. Genet. Dev.* **10**, 81–93.

Marmorstein, L. Y., Ouchi, T., and Aaronson, S. A. (1998). The BRCA2 gene product functionally interacts with p53 and RAD51. *Proc. Natl. Acad. Sci. USA* **95**, 13869–13874.

Mazzarella, R., and Schlessinger, D. (1998). Pathological consequences of sequence duplications in the human genome. *Genome Res.* **8**, 1007–1021.

Mekeel, K. L., Tang, W., Kachnic, L. A., Luo, C. M., DeFrank, J. S., and Powell, S. N. (1997). Inactivation of p53 results in high rates of homologous recombination. *Oncogene* **14**, 1847–1857.

Meyn, M. S. (1999). Ataxia-telangiectasia, cancer and the pathobiology of the ATM gene. *Clin. Genet.* **55**, 289–304.

Mitelman, F., Johansson, B., and Mertens, F. (2003). Mitelman Database of Chromosomal Aberrations in Cancer. http://cgap.nci.nih.gov/Chromosomes/Mitelman.

Moynahan, M. E., Pierce, A. J., and Jasin, M. (2001). BRCA2 is required for homology-directed repair of chromosomal breaks. *Mol. Cell* **7**, 263–272.

Murphy, K. M., Brune, K. A., Griffin, C., Sollenberger, J. E., Petersen, G. M., Bansal, R., Hruban, R. H., and Kern, S. E. (2002). Evaluation of candidate genes MAP2K4, MADH4, ACVR1B, and BRCA2 in familial pancreatic cancer: Deleterious BRCA2 mutations in 17%. *Cancer Res.* **62**, 3789–3793.

Nakanishi, K., Yang, Y. G., Pierce, A. J., Taniguchi, T., Digweed, M., D'Andrea, A. D., Wang, Z. Q., and Jasin, M. (2005). Human Fanconi anemia monoubiquitination pathway promotes homologous DNA repair. *Proc. Natl. Acad. Sci. USA* **102**, 1110–1115.

Nathanson, K. L., Wooster, R., Weber, B. L., and Nathanson, K. N. (2001). Breast cancer genetics: What we know and what we need. *Nat. Med.* **7**, 552–556.

Naylor, J., Brinke, A., Hassock, S., Green, P. M., and Giannelli, F. (1993). Characteristic mRNA abnormality found in half the patients with severe haemophilia A is due to large DNA inversions. *Hum. Mol. Genet.* **2**, 1773–1778.

Naylor, J. A., Nicholson, P., Goodeve, A., Hassock, S., Peake, I., and Giannelli, F. (1996). A novel DNA inversion causing severe hemophilia A. *Blood* **87**, 3255–3261.

Nelis, E., Van Broeckhoven, C., De Jonghe, P., Lofgren, A., Vandenberghe, A., Latour, P., Le Guern, E., Brice, A., Mostacciuolo, M. L., Schiavon, F., *et al.* (1996). Estimation of the mutation frequencies in Charcot-Marie-Tooth disease type 1 and hereditary neuropathy with liability to pressure palsies: A European collaborative study. *Eur. J. Hum. Genet.* **4**, 25–33.

Oshima, J., Huang, S., Pae, C., Campisi, J., and Schiestl, R. H. (2002). Lack of WRN results in extensive deletion at nonhomologous joining ends. *Cancer Res.* **62**, 547–551.

Pentao, L., Wise, C. A., Chinault, A. C., Patel, P. I., and Lupski, J. R. (1992). Charcot-Marie-Tooth type 1A duplication appears to arise from recombination at repeat sequences flanking the 1.5 Mb monomer unit. *Nat. Genet.* **2**, 292–300.

Pierce, A. J., and Jasin, M. (2001). NHEJ deficiency and disease. *Mol. Cell* **8**, 1160–1161.

Pittman, D. L., and Schimenti, J. C. (2000). Midgestation lethality in mice deficient for the RecA-related gene, Rad51d/Rad51l3. *Genesis* **26**, 167–173.

Purandare, S. M., and Patel, P. I. (1997). Recombination hot spots and human disease. *Genome Res.* **7**, 773–786.

Rafii, S., O'Regan, P., Xinarianos, G., Azmy, I., Stephenson, T., Reed, M., Meuth, M., Thacker, J., and Cox, A. (2002). A potential role for the XRCC2 R188H polymorphic site in DNA-damage repair and breast cancer. *Hum. Mol. Genet.* **11**, 1433–1438.

Reichenbach, J., Schubert, R., Schwan, C., Muller, K., Bohles, H. J., and Zielen, S. (1999). Anti-oxidative capacity in patients with ataxia telangiectasia. *Clin. Exp. Immunol.* **117,** 535–539.

Reichenbach, J., Schubert, R., Schindler, D., Muller, K., Bohles, H., and Zielen, S. (2002). Elevated oxidative stress in patients with ataxia telangiectasia. *Antioxid. Redox Signal.* **4,** 465–469.

Reliene, R., and Schiestl, R. H. (2003). Mouse models for induced genetic instability at endogenous loci. *Oncogene* **22,** 7000–7010.

Reliene, R., Fischer, E., and Schiestl, R. H. (2004). Effect of N-acetyl cysteine on oxidative DNA damage and the frequency of DNA deletions in atm-deficient mice. *Cancer Res.* **64,** 5148–5153.

Reliene, R., Hlavacova, A., Mahadevan, B., Baird, W. M., and Schiestl, R. H. (2005). Diesel exhaust particles cause increased levels of DNA deletions after transplacental exposure in mice. *Mutat. Res.* **570,** 245–252.

Richardson, C., Stark, J. M., Ommundsen, M., and Jasin, M. (2004). Rad51 overexpression promotes alternative double-strand break repair pathways and genome instability. *Oncogene* **23,** 546–553.

Rodriguez-Lopez, R., Osorio, A., Ribas, G., Pollan, M., Sanchez-Pulido, L., de la Hoya, M., Ruibal, A., Zamora, P., Arias, J. I., Salazar, R., *et al.* (2004). The variant E233G of the RAD51D gene could be a low-penetrance allele in high-risk breast cancer families without BRCA1/2 mutations. *Int. J. Cancer* **110,** 845–849.

Schichman, S. A., Caligiuri, M. A., Strout, M. P., Carter, S. L., Gu, Y., Canaani, E., Bloomfield, C. D., and Croce, C. M. (1994). ALL-1 tandem duplication in acute myeloid leukemia with a normal karyotype involves homologous recombination between Alu elements. *Cancer Res.* **54,** 4277–4280.

Schiestl, R. H., Igarashi, S., and Hastings, P. J. (1988). Analysis of the mechanism for reversion of a disrupted gene. *Genetics* **119,** 237–247.

Schiestl, R. H., Gietz, R. D., Mehta, R. D., and Hastings, P. J. (1989a). Carcinogens induce intrachromosomal recombination in yeast. *Carcinogenesis* **10,** 1445–1455.

Schiestl, R. H., Chan, W. S., Gietz, R. D., Mehta, R. D., and Hastings, P. J. (1989b). Safrole, eugenol and methyleugenol induce intrachromosomal recombination in yeast. *Mutat. Res.* **224,** 427–436.

Schiestl, R. H., Khogali, F., and Carls, N. (1994). Reversion of the mouse pink-eyed unstable mutation induced by low doses of x-rays. *Science* **266,** 1573–1576.

Schiestl, R. H., Aubrecht, J., Yap, W. Y., Kandikonda, S., and Sidhom, S. (1997a). Polychlorinated biphenyls and 2,3,7,8-tetrachlorodibenzo-p-dioxin induce intrachromosomal recombination *in vitro* and *in vivo*. *Cancer Res.* **57,** 4378–4383.

Schiestl, R. H., Aubrecht, J., Khogali, F., and Carls, N. (1997b). Carcinogens induce reversion of the mouse pink-eyed unstable mutation. *Proc. Natl. Acad. Sci. USA* **94,** 4576–4581.

Schmid, C. W., Deka, N., and Matera, A. G. (1989). Repetitive human DNA: The shape of things to come. *In* "Chromosomes: Eucaryotic, Procaryotic and Viral" (K. W. Adolph, ed.), Vol. I, pp. 3–29. CRC Press, Boca Raton.

Scully, R., and Livingston, D. M. (2000). In search of the tumour-suppressor functions of BRCA1 and BRCA2. *Nature* **408,** 429–432.

Shen, J., and Loeb, L. A. (2001). Unwinding the molecular basis of the Werner syndrome. *Mech. Ageing Dev.* **122,** 921–944.

Shivji, M. K., and Venkitaraman, A. R. (2004). DNA recombination, chromosomal stability and carcinogenesis: Insights into the role of BRCA2. *DNA Repair (Amst.)* **3,** 835–843.

Shu, Z., Smith, S., Wang, L., Rice, M. C., and Kmiec, E. B. (1999). Disruption of muREC2/RAD51L1 in mice results in early embryonic lethality which can be partially rescued in a p53 (−/−) background. *Mol. Cell. Biol.* **19,** 8686–8693.

Snouwaert, J. N., Gowen, L. C., Latour, A. M., Mohn, A. R., Xiao, A., DiBiase, L., and Koller, B. H. (1999). BRCA1 deficient embryonic stem cells display a decreased homologous recombination frequency and an increased frequency of non-homologous recombination that is corrected by expression of a brca1 transgene. *Oncogene* **18**, 7900–7907.

Strong, L. C., Williams, W. R., and Tainsky, M. A. (1992). The Li-Fraumeni syndrome: From clinical epidemiology to molecular genetics. *Am. J. Epidemiol.* **135**, 190–199.

Stumm, M., Neubauer, S., Keindorff, S., Wegner, R. D., Wieacker, P., and Sauer, R. (2001). High frequency of spontaneous translocations revealed by FISH in cells from patients with the cancer-prone syndromes ataxia telangiectasia and Nijmegen breakage syndrome. *Cytogenet. Cell. Genet.* **92**, 186–191.

Sturzbecher, H. W., Donzelmann, B., Henning, W., Knippschild, U., and Buchhop, S. (1996). p53 is linked directly to homologous recombination processes via RAD51/RecA protein interaction. *EMBO J.* **15**, 1992–2002.

Taniguchi, T., Garcia-Higuera, I., Andreassen, P. R., Gregory, R. C., Grompe, M., and D'Andrea, A. D. (2002). S-phase-specific interaction of the Fanconi anemia protein, FANCD2, with BRCA1 and RAD51. *Blood* **100**, 2414–2420.

Thacker, J. (2005). The RAD51 gene family, genetic instability and cancer. *Cancer Lett.* **219**, 125–135.

Thompson, L. H., and Schild, D. (2001). Homologous recombinational repair of DNA ensures mammalian chromosome stability. *Mutat. Res.* **477**, 131–153.

Thompson, L. H., and Schild, D. (2002). Recombinational DNA repair and human disease. *Mutat. Res.* **509**, 49–78.

Thompson, L. H., Hinz, J. M., Yamada, N. A., and Jones, N. J. (2005). How Fanconi anemia proteins promote the four Rs: Replication, recombination, repair, and recovery. *Environ. Mol. Mutagen.* **45**, 128–142.

Todd, R., and Wong, D. T. (1999). Oncogenes. *Anticancer Res.* **19**, 4729–4746.

Tsuzuki, T., Fujii, Y., Sakumi, K., Tominaga, Y., Nakao, K., Sekiguchi, M., Matsushiro, A., Yoshimura, Y., and Morita, T. (1996). Targeted disruption of the Rad51 gene leads to lethality in embryonic mice. *Proc. Natl. Acad. Sci. USA* **93**, 6236–6240.

Tutt, A., Gabriel, A., Bertwistle, D., Connor, F., Paterson, H., Peacock, J., Ross, G., and Ashworth, A. (1999). Absence of Brca2 causes genome instability by chromosome breakage and loss associated with centrosome amplification. *Curr. Biol.* **9**, 1107–1110.

Uziel, T., Lerenthal, Y., Moyal, L., Andegeko, Y., Mittelman, L., and Shiloh, Y. (2003). Requirement of the MRN complex for ATM activation by DNA damage. *EMBO J.* **22**, 5612–5621.

Venkitaraman, A. R. (2001). Chromosome stability, DNA recombination and the BRCA2 tumour suppressor. *Curr. Opin. Cell. Biol.* **13**, 338–343.

Vispe, S., Cazaux, C., Lesca, C., and Defais, M. (1998). Overexpression of Rad51 protein stimulates homologous recombination and increases resistance of mammalian cells to ionizing radiation. *Nucleic Acids Res.* **26**, 2859–2864.

Wang, W., Seki, M., Narita, Y., Sonoda, E., Takeda, S., Yamada, K., Masuko, T., Katada, T., and Enomoto, T. (2000a). Possible association of BLM in decreasing DNA double strand breaks during DNA replication. *EMBO J.* **19**, 3428–3435.

Wang, X. W., Tseng, A., Ellis, N. A., Spillare, E. A., Linke, S. P., Robles, A. I., Seker, H., Yang, Q., Hu, P., Beresten, S., *et al.* (2001). Functional interaction of p53 and BLM DNA helicase in apoptosis. *J. Biol. Chem.* **276**, 32948–32955.

Wang, Y., Cortez, D., Yazdi, P., Neff, N., Elledge, S. J., and Qin, J. (2000b). BASC, a super complex of BRCA1-associated proteins involved in the recognition and repair of aberrant DNA structures. *Genes Dev.* **14**, 927–939.

Wijnhoven, S. W., Kool, H. J., van Teijlingen, C. M., van Zeeland, A. A., and Vrieling, H. (2001). Loss of heterozygosity in somatic cells of the mouse. An important step in cancer initiation? *Mutat. Res.* **473**, 23–36.

Wise, C. A., Garcia, C. A., Davis, S. N., Heju, Z., Pentao, L., Patel, P. I., and Lupski, J. R. (1993). Molecular analyses of unrelated Charcot-Marie-Tooth (CMT) disease patients suggest a high frequency of the CMTIA duplication. *Am. J. Hum. Genet.* **53,** 853–863.

Wong, A. K., Pero, R., Ormonde, P. A., Tavtigian, S. V., and Bartel, P. L. (1997). RAD51 interacts with the evolutionarily conserved BRC motifs in the human breast cancer susceptibility gene brca2. *J. Biol. Chem.* **272,** 31941–31944.

Wu, L., Davies, S. L., Levitt, N. C., and Hickson, I. D. (2001). Potential role for the BLM helicase in recombinational repair via a conserved interaction with RAD51. *J. Biol. Chem.* **276,** 19375–19381.

Xu, X., Weaver, Z., Linke, S. P., Li, C., Gotay, J., Wang, X. W., Harris, C. C., Ried, T., and Deng, C. X. (1999). Centrosome amplification and a defective G2-M cell cycle checkpoint induce genetic instability in BRCA1 exon 11 isoform-deficient cells. *Mol. Cell* **3,** 389–395.

Yen, P. H., Li, X. M., Tsai, S. P., Johnson, C., Mohandas, T., and Shapiro, L. J. (1990). Frequent deletions of the human X chromosome distal short arm result from recombination between low copy repetitive elements. *Cell* **61,** 603–610.

Yin, Y., Tainsky, M. A., Bischoff, F. Z., Strong, L. C., and Wahl, G. M. (1992). Wild-type p53 restores cell cycle control and inhibits gene amplification in cells with mutant p53 alleles. *Cell* **70,** 937–948.

Yoshikawa, K., Ogawa, T., Baer, R., Hemmi, H., Honda, K., Yamauchi, A., Inamoto, T., Ko, K., Yazumi, S., Motoda, H., *et al.* (2000). Abnormal expression of BRCA1 and BRCA1-interactive DNA-repair proteins in breast carcinomas. *Int. J. Cancer* **88,** 28–36.

Zhang, H., Somasundaram, K., Peng, Y., Tian, H., Bi, D., Weber, B. L., and El-Deiry, W. S. (1998). BRCA1 physically associates with p53 and stimulates its transcriptional activity. *Oncogene* **16,** 1713–1721.

Zhou, Z. H., Akgun, E., and Jasin, M. (2001). Repeat expansion by homologous recombination in the mouse germ line at palindromic sequences. *Proc. Natl. Acad. Sci. USA* **98,** 8326–8333.

4

Mutational Analysis of the Ribosome

Kathleen L. Triman
Department of Biology, Franklin and Marshall College
Lancaster, Pennsylvania 17604

Advances in Genetics, Vol. 58
Copyright 2007, Elsevier Inc. All rights reserved.

0065-2660/07 $35.00
DOI: 10.1016/S0065-2660(06)58004-6

ABSTRACT

The ribosome is responsible for protein synthesis, the translation of the genetic code, in all living organisms. Ribosomes are composed of RNA (ribosomal RNA) and protein (ribosomal protein). Soluble protein factors bind to the ribosome and facilitate different phases of translation. Genetic approaches have proved useful for the identification and characterization of the structural and functional roles of specific nucleotides in ribosomal RNA and of specific amino acids in ribosomal proteins and in ribosomal factors. This chapter summarizes examples of mutations identified in ribosomal RNA, ribosomal proteins, and ribosomal factors.   © 2007, Elsevier Inc.

I. INTRODUCTION

Ribosomes are found in all cells and are the sites of translation of genetic information into protein. Each ribosome is composed of a large and a small subunit. The simplest ribosomes (those from bacteria and archaea) contain about 50 different proteins and 3 ribosomal RNAs (16S, 23S, and 5S rRNAs) comprising about 4500 nucleotides and 60% of the mass of the ribosome. In addition to the ribosomal proteins, many nonribosomal protein factors are required for the initiation, elongation, and termination steps of translation. Anders Liljas provides an excellent text containing a summary of the structural knowledge of the translation system (Liljas, 2004).

The purpose of this chapter is to summarize the vast and rapidly growing literature on mutational analysis of the ribosome, including examples of analyses for rRNA, ribosomal proteins, and ribosomal factors. Previous reviews were more narrowly focused on mutational analysis of 16S rRNA (Triman, 1995) and mutational analysis of 23S rRNA (Triman, 1999).

Remarkable progress in the determination of structures of ribosomal subunits at atomic resolution (summarized on p. 11 in Liljas, 2004) and of whole ribosomes (Schuwirth et al., 2005) has completely changed the field of protein synthesis (reviewed in Moore, 2005; Moore and Steitz, 2005; Noller, 2005). Information provided by biochemical, enzymological, genetic, and structural studies establishes that the active site of the large ribosomal subunit, where peptide bond formation occurs, is composed entirely of the RNA (Moore and Steitz, 2006; Noller, 2006). Studies provide evidence for a precise positioning of active site nucleotides due to a direct Watson–Crick interaction between C75 of aminoacyl-tRNA and G2553 of the 23S rRNA (Brunelle et al., 2006; Cochella and Green, 2005 and references therein). Mutational analysis of the ribosome now includes an exciting new category of ribosome variants, those altered at specific aminoacyl-tRNA nucleotides that can suppress rRNA mutations (Dorner et al., 2006).

II. MUTATIONAL ANALYSIS OF 16S rRNA STRUCTURE AND FUNCTION

The 16S rRNA molecule is subdivided into three major structural domains and one minor domain by three sets of long-range base-paired interactions [refer to Triman (1995) and to updated secondary structures in Cannone et al. (2002)]. Examples of the results of mutational analysis of 16S rRNA structure and function presented in this chapter are organized according to the structural domain in which particular mutations are located and at positions from the 5' to the 3' end of 16S rRNA corresponding to equivalent nucleotide positions in *Escherichia coli*.

A. Mutations in the 5' major domain of 16S rRNA

1. Deletion of U12 and U12C substitution correct the subunit association and translational fidelity defects caused by the A900G mutation (Belanger et al., 2005).
2. A random mutant library of rRNA genes was prepared and dominant mutations that interfered with cell growth were selected. Fifty-three 16S rRNA mutations were identified and ranked according to the severity of the phenotype. Twelve of the 53 mutations are clustered around sites targeted by well-characterized antibiotics. Known drugs do not target the rest. This study provides potential targets that may prove useful in the identification of new antibiotic targets. Among these mutations are C18G, U49C, A51G, A55G, G57A, A161G, G299A, A373G, and A389G (Yassin et al., 2005).
3. A pyrimidine (Psi or C) is required at position 516. The authors find no evidence for base pair formation between Psi516 and A535 (Lee et al., 2001).
4. A random mutant library of rRNA genes was prepared and dominant mutations that interfered with cell growth were selected. Fifty-three 16S rRNA mutations were identified and ranked according to the severity of the phenotype. Twelve of the 53 mutations are clustered around sites targeted by well-characterized antibiotics. Known drugs do not target the rest. This approach is useful in the identification of new antibiotic targets. Among these mutations is G521A (Yassin et al., 2005).
5. A novel mutation G524C confers streptomycin resistance in *Mycobacterium smegmatis*. C526U, C522U, A523C also conferred resistance (Springer et al., 2001).
6. Any substitution at G530 abolished translation activity *in vivo* (Abdi and Fredrick, 2005).
7. Substitutions at position 535 reduce ribosomal function by 50%. The authors find no evidence for base pair formation between Psi516 and A535 (Lee et al., 2001).

B. Mutations in the central domain of 16S rRNA

1. A random mutant library of rRNA genes was prepared and dominant mutations that interfered with cell growth were selected. Fifty-three 16S rRNA mutations were identified and ranked according to the severity of the phenotype. Twelve of the 53 mutations are clustered around sites targeted by well-characterized antibiotics. Known drugs do not target the rest. This approach is useful in the identification of new antibiotic targets. Among these mutations are C614A and A622G (Yassin et al., 2005).

2. Random mutagenesis of positions 690–697 produced 101 unique functional mutants. Double mutants containing G690 U697 and G691 A696 formed sheared hydrogen-bonded mismatches. Hyperactivity compared to wild type was observed in single mutations at 693. Nucleotide identity significantly affected ribosomal function at 690, 695, 696, and 697. Pyrimidines were absent at 696, as were C at 691, G at 697. Covariation between 690 and 697 was observed (Morosyuk et al., 2001).

3. A694G confers pactamycin resistance. Ribosomal functions were not affected by this mutation (Mankin, 1997).

4. Mutations at C732 alter the cellular responses to decreases in Ffh (bacterial homologue of the mammalian signal recognition particle protein) and 4.5S RNA (bacterial homologue of the mammalian signal recognition particle RNA) levels (Brunelli et al., 2002).

5. 190 of the possible 262,144 possible mutant sequences in the 790 loop between positions 786 and 796 were functional. Positions 789 and 791 were invariant in functional mutants. Covariation observed at positions 787, 788, 794, and 795. Weak pairing interactions between 787 and 795 and between 788 and 794. Stable pairing between 787 and 795 prevents subunit association (Lee et al., 1997).

6. All base changes at A794 confer resistance to kasugamycin; only one mutation, A794C, could not be constructed (Vila-Sanjurjo et al., 1999).

7. C795U and C796U confer pactamycin resistance. Ribosomal functions were not affected by these mutations (Mankin, 1997).

8. A random mutant library of rRNA genes was prepared and dominant mutations that interfered with cell growth were selected. Fifty-three 16S rRNA mutations were identified and ranked according to the severity of the phenotype. Twelve of the 53 mutations are clustered around sites targeted by well-characterized antibiotics. Known drugs do not target the rest. This approach is useful in the identification of new antibiotic targets. Among these mutations are A802G and U804C (Yassin et al., 2005).

9. Quadruple mutation, C912G, C910G, G885C, and G887C, is deleterious in the presence of the A2058G marker in 23S rRNA and C1192U marker in 16S rRNA. However, the quadruple mutant grows at wild-type rates in the absence of these markers. Thus, the markers are not phenotypically silent (Rodriguez-Correa and Dahlberg, 2004).
10. Randomized four nucleotides in the tetraloop GNRA motif at positions 898–901. Positions 899 and 900 were found to be the most critical for ribosome activity. Functional studies showed that mutations in the 900 tetraloop impair subunit association and decrease translational fidelity (Belanger et al., 2002).
11. A900G severely impairs subunit association and translational fidelity. Deletion of U12 and U12C substitution correct the subunit association and translational fidelity defects caused by the A900G mutation (Belanger et al., 2005).
12. Quadruple mutation, C912G, C910G, G885C, and G887C, is deleterious in the presence of the A2058G marker in 23S rRNA and C1192U marker in 16S rRNA. However, the quadruple mutant grows at wild-type rates in the absence of these markers. Thus, the markers are not phenotypically silent (Rodriguez-Correa and Dahlberg, 2004).

C. Mutations in the 3' major domain of 16S rRNA

1. All base changes at G926 confer resistance to kasugamycin (Vila-Sanjurjo et al., 1999).
2. Deletion of G942 confers very weak tetracycline resistance in *Helicobacter pylori* (Trieber and Taylor, 2002).
3. A random mutant library of rRNA genes was prepared and dominant mutations that interfered with cell growth were selected. Fifty-three 16S rRNA mutations were identified and ranked according to the severity of the phenotype. Twelve of the 53 mutations are clustered around sites targeted by well-characterized antibiotics. Known drugs do not target the rest. This approach is useful in the identification of new antibiotic targets. Among these mutations is A964G (Yassin et al., 2005).
4. The triple mutation AGA965–967UUC confers tetracycline resistance in *H. pylori* (Trieber and Taylor, 2002).
5. Double mutations in the primary tetracycline-binding site AGA at 965–967 include gGA, AGc, guA, or gGc (lowercase letters indicate base changes) and resulted in tetracycline resistance (Dailidiene et al., 2002).
6. G966A decreased translation activity *in vivo*. G1338A suppresses phenotypes conferred by G966A (Abdi and Fredrick, 2005).

7. C1054A caused a defect in catalysis of peptidyl-tRNA hydrolysis during peptide chain termination in the presence of release factor (RF)2 (Arkov *et al.*, 1998).

8. G1058C confers resistance to tetracycline in gram-positive cutaneous proprionibacteria (Ross *et al.*, 1998).

9. A random mutant library of rRNA genes was prepared and dominant mutations that interfered with cell growth were selected. Fifty-three 16S rRNA mutations were identified and ranked according to the severity of the phenotype. Twelve of the 53 mutations are clustered around sites targeted by well-characterized antibiotics. Known drugs do not target the rest. This approach is useful in the identification of new antibiotic targets. Among these mutations is G1058A (Yassin *et al.*, 2005).

10. C1066U affects the requirement for 4.5S RNA (bacterial homologue of the mammalian signal recognition particle RNA) (Brunelli *et al.*, 2002).

11. A random mutant library of rRNA genes was prepared and dominant mutations that interfered with cell growth were selected. Fifty-three 16S rRNA mutations were identified and ranked according to the severity of the phenotype. Twelve of the 53 mutations are clustered around sites targeted by well-characterized antibiotics. Known drugs do not target the rest. This approach is useful in the identification of new antibiotic targets. Among these mutations are G1068A, G1072A, U1073C, U1085C, and A1111U (Yassin *et al.*, 2005).

12. A1191G and C1192U each confers high-level resistance to spectinomycin in *Borrelia burgdorferi* (Criswell *et al.*, 2006).

13. C1192U is deleterious in the presence of an additional U1406A mutation. C1192U enhances the drug resistance conferred by A1408G (Recht and Puglisi, 2001).

14. The C1192U marker in 16S rRNA is not phenotypically silent. Quadruple mutation, C912G, C910G, G885C, and G887C, is deleterious in the presence of the A2058G marker in 23S rRNA and C1192U marker in 16S rRNA. However, the quadruple mutant grows at wild-type rates in the absence of these markers (Rodriguez-Correa and Dahlberg, 2004).

15. Alterations at G1338 or A1339 did not completely inactivate the ribosome. A >18-fold decrease in translation activity *in vivo* is conferred by any mutation at A1339. G1338A suppresses phenotypes conferred by G966A and results in highly active ribosomes. G1338C and G1338U decreased translation by 11- and 4.5-fold (Abdi and Fredrick, 2005).

16. Decreases in translation were observed after mutagenesis of G1338 and A1339. G1338A suppresses the effects of A1339G and of the double mutant A790C C1400U. The addition of 50S subunits suppresses the effects of mutations at 1338 and 1339 and the effect of the double mutation G1338C:A1339C (Lancaster and Noller, 2005).

D. Mutations in the 3′ minor domain of 16S rRNA

1. A random mutant library of rRNA genes was prepared and dominant mutations that interfered with cell growth were selected. Fifty-three 16S rRNA mutations were identified and ranked according to the severity of the phenotype. Twelve of the 53 mutations are clustered around sites targeted by well-characterized antibiotics. Known drugs do not target the rest. This approach is useful in the identification of new antibiotic targets. Among these mutations is C1395U (Yassin *et al.*, 2005).

2. A1400G, C1401U, and double mutation C1401A with G1483U confer resistance to kanamycin in M. *smegmatis* and M. *tuberculosis* (Suzuki *et al.*, 1998).

3. All possible mutations at C1402 were constructed; one single mutant, C1402U, and five double mutants, C1402U with A1500U, C1402A with A1500U, C1402U with A1500G, C1402U with A1500C, and C1402A with A1500G, were viable. A suppressor mutation specific for C1402U A1500G was also isolated at 1520: C1520U (Vila-Sanjurjo and Dahlberg, 2001).

4. U1406A confers resistance to kanamycin A and G418 and moderate resistance to gentamicin C and tobramycin when expressed in a homogeneous population of ribosomes. U1406A is deleterious in the presence of an additional C1192U mutation (Recht and Puglisi, 2001).

5. U1406A, C, or G confers hygromycin B resistance in *Thermus thermophilus* (Gregory *et al.*, 2005a).

6. U1406C confers hygromycin B resistance in M. *smegmatis* (Pfister *et al.*, 2003b).

7. A1408G confers high-level resistance to kanamycin A, neomycin, and gentamicin and moderate resistance to paramomycin when expressed in a homogeneous population of ribosomes. When expressed at levels greater than 50% of the ribosome population, A1408G resistance is enhanced by an additional C1192U mutation (Recht and Puglisi, 2001).

8. A1408G confers high-level resistance to kanamycin and gentamicin in B. *burgdorferi* (Criswell *et al.*, 2006).

9. A1408G confers streptomycin resistance (Gregory *et al.*, 2005a), influences conformational changes during tRNA selection, and results in kanamycin and capreomycin resistance in T. *thermophilus* (Gregory *et al.*, 2005b).

10. Two regions appear crucial for binding in the A site: A1408 and the non-Watson–Crick U1406 U1495 pair. U1406C U1495C and U1406A U1495A double mutants confer low to moderate levels of drug resistance, whereas U1406C/U1495A confers high-level resistance to aminoglycosides (except for neomycin) (Pfister *et al.*, 2003a).

11. M. *abscessus*, M. *chelonae*, and M. *smegmatis* amikacin resistance mutations resulted from A1408G (Prammananan *et al.*, 1998).

12. M. *smegmatis* mutants with altered bases at 1409 and 1491 discriminated poorly between disubstituted streptamines; G1491U conferred high-level resistance to paramomycin and geneticin, but not to neomycin, tobramycin, or gentamicin. In contrast to *E. coli*, C1409G and G1491A were viable in M. *smegmatis*. C1409U and G1491U did not affect growth rate. Growth retardation is associated with C1409G, G1491A, and G1491C (Pfister *et al.*, 2004a).

13. C1409G confers capreomycin and streptomycin resistance in *T. thermophilus* (Gregory *et al.*, 2005b).

14. G1416 is one of 44 new positions among 59 mutated positions in a large-scale combinatorial mutagenesis and *in vivo* selection carried out on 16S rRNA nucleotides that form RNA-RNA intersubunit bridges in the *E. coli* ribosome. Mutation of G1416 reduces subunit association (Rackham *et al.*, 2006).

15. Antidownstream box flip mutations collectively reverse the orientation of the 24 base-paired nucleotides in the 12 base pair penultimate stem (helix 44) between 1419::1481 and 1430::1470. For example, G1419U and U1481G substitutions, and so on, are made. The mutants are indistinguishable from wild type with respect to cold acclimation, the ability to withstand cold shock, and the efficiency of translation of mRNAs that contain or lack downstream box sequences (LaTeana *et al.*, 2001).

16. G1491A confers capreomycin resistance in *T. thermophilus* (Gregory *et al.*, 2005b).

17. Any substitution at A1492 abolished translation activity *in vivo*. A1493C abolished translation activity, whereas A1493G and A1493U retained residual translation activity *in vivo* (Abdi and Fredrick, 2005).

18. U1495C confers hygromycin B resistance in *T. thermophilus* (Gregory *et al.*, 2005b).

19. M. *smegmatis* resistance mutations, C1496U and U1498C, confer hygromycin B resistance (Pfister *et al.*, 2003a).

20. All possible mutations at A1500 were constructed; one single mutant, A1500U, and five double mutants, C1402U with A1500U, C1402A with A1500U, C1402U with A1500G, C1402U with A1500C, and C1402A with A1500G, were viable. A suppressor mutation specific for C1402U A1500G was also isolated at 1520: C1520U (Vila-Sanjurjo and Dahlberg, 2001).

21. All base changes at A1519 confer resistance to kasugamycin. A1518 mutations were deleterious and/or impaired ribosomal function (Vila-Sanjurjo *et al.*, 1999).

22. C1520U suppresses double mutation C1402U A1500G (Vila-Sanjurjo and Dahlberg, 2001).

III. MUTATIONAL ANALYSIS OF 23S rRNA STRUCTURE AND FUNCTION

The 23S rRNA molecule is subdivided into six major structural domains by long-range base-paired interactions [refer to Triman (1999) and to updated secondary structures in Cannone *et al.* (2002)]. Examples of the results of mutational analysis of 23S rRNA structure and function presented in this chapter are organized according to the structural domain in which particular mutations are located and at positions from the 5' to the 3' end of 23S rRNA corresponding to equivalent nucleotide positions in *E. coli*.

A. Mutations in domain I of 23S rRNA

A64U, U65A, C66G; A64U, U65A, C66G/G88C, A89U, U90A; UUCG replacement of nucleotides 59–68 or 88–94: four constructs produced changes to disrupt pairing in hairpins proposed to produce a pseudoknot structure. The results rule out the functionality of the proposed pseudoknot *in vivo* (Chernyaeva and Murgola, 2000).

B. Mutations in domains II–IV of 23S rRNA

1. Methylation at G748: single methylations at 23S rRNA nucleotides G748 and A2058 acting in synergy confer resistance to the macrolide antibiotic tylosin (Liu and Douthwaite, 2002).
2. U754A confers ketolide resistance (Xiong *et al.*, 1999).
3. U754A reduces tryptophan induction of tna operon expression (Cruz-Vera *et al.*, 2005).
4. A960C, A960G, A960U: mutations influence the structure of 5S rRNA and the peptidyl transferase region of 23S rRNA (Sergiev *et al.*, 2000).
5. A1067U: extensive deletion analysis demonstrated that base-pairing of 23S rRNA ends is essential for ribosomal large-subunit assembly (Liiv and Remme, 1998).
6. Deletion and substitution at 1067, 1093, 1094, 1095, A1098C: mutations indicate RF2-interactive sites (Xu *et al.*, 2002).
7. G1093A: mutational evidence for a functional connection between two domains of 23S rRNA in translation termination (Arkov *et al.*, 2002).
8. A1095G, A1095C: thiostrepton-resistant mutants of *T. thermophilus* (Cameron *et al.*, 2004).
9. Psi1911C mutation does not affect ribosomal function (Liiv *et al.*, 2005).
10. U1915A results in a severe growth defect and low translational fidelity (Hirabayashi *et al.*, 2006).

C. Mutations in domain V of 23S rRNA

1. A2030G confers chloramphenicol resistance in *T. thermophilus* (Gregory *et al.*, 2005a).
2. G2032A, U, or C confers oxazolidinone resistance (Xiong *et al.*, 2000).
3. G2032A, G2447A: catalytic properties of mutant 23S ribosomes resistant to oxazolidinones (Bobkova *et al.*, 2003).
4. Deletion of CG from CGCG 2045–2048: mutational evidence for a functional connection between two domains of 23S rRNA in translation termination (Arkov *et al.*, 2002).
5. A2058G or A2058U confers resistance to azithromycin and erythromycin in *Staphylococcus aureus* (Prunier *et al.*, 2002).
6. 2058A > G confers macrolide resistance mutation (Pfister *et al.*, 2005).
7. Resistance to the macrolide antibiotic tylosin is conferred by single methylations at 23S rRNA nucleotides G748 and A2058 acting in synergy (Liu and Douthwaite, 2002).
8. A2058G-A2059G-A2062G: mutation in domain V of the 23S rRNA of *Bacillus subtilis* that inactivates its protein-folding property *in vitro* (Chowdhury *et al.*, 2002).
9. A2058G: binding site of the bridged macrolides in the *E. coli* ribosome (Xiong *et al.*, 2005).
10. A2058G: spontaneous erythromycin-resistant mutation *T. thermophilus* IB-21 (Gregory *et al.*, 2001c).
11. A2058G, A2059G: mutations in the 23S rRNA gene of *H. pylori* associated with clarithromycin resistance (Kim *et al.*, 2002).
12. Methylation at A2058: base-pairing of 23S rRNA ends is essential for ribosomal large-subunit assembly (Liiv and Remme, 1998).
13. A2058G, A2059G: transition mutations in the 23S rRNA of erythromycin-resistant isolates of *Mycoplasma pneumoniae* (Lucier *et al.*, 1995).
14. A2058G with either C1192 or C1400U in 16S rRNA: double mutants suppress temperature-sensitive EF-G mutation (Koosha *et al.*, 2000).
15. Quadruple 16S rRNA mutation, C912G, C910G, G885C, and G887C, is deleterious in the presence of the A2058G marker in 23S rRNA and C1192U marker in 16S rRNA. However, the quadruple mutant grows at wild-type rates in the absence of these markers. Thus, the markers are not phenotypically silent (Rodriguez-Correa and Dahlberg, 2004).
16. A2058 to G, C, or U; A2059G, C; A2058C-A2059C: structural basis of macrolide–ribosome binding assessed (Pfister *et al.*, 2004b).
17. 2058, 2059: molecular basis of clarithromycin resistance in M. *avium intracellulare* complex (Jamal *et al.*, 2000).

18. A2058, A2059: mutations confer resistance to clindamycin in M. *smegmatis* (Poehlsgaard *et al.*, 2005).
19. A2058G, U2609C: binding site of the bridged macrolides in the E. *coli* ribosome (Xiong *et al.*, 2005).
20. A2059G confers resistance to azithromycin and erythromycin in S. *aureus* (Prunier *et al.*, 2002).
21. A2059G confers tylosin resistance in T. *thermophilus* (Gregory *et al.*, 2005a).
22. A2060G mutation confers chloramphenicol resistance in mitochondria of Chinese hamster cells (Hashiguchi and Ikushima, 2000).
23. G2061U confers clindamycin resistance in *Toxoplasma gondii* (Camps *et al.*, 2002).
24. G2061A confers chloramphenicol resistance in T. *thermophilus* (Gregory *et al.*, 2005a).
25. A2058G-A2059G-A2062G: mutations in domain V of the 23S rRNA of B. *subtilis* that inactivate its protein-folding property *in vitro* (Chowdhury *et al.*, 2002).
26. A2058U, A2062G, A2062C, A2058U-A2062G: antibiotic resistance mutations isolated using an *in vitro* selection system (Cochella and Green, 2004).
27. A2062C: mutation responsible for resistance to 16-membered macrolides and streptogramins in *Streptococcus pneumoniae* (Depardieu and Courvalin, 2001).
28. A2062G, A2062U: mutations associated with resistance to 16-membered macrolide antibiotics in M. *hominis* (Furneri *et al.*, 2001).
29. A2062G confers tylosin and chloramphenicol resistance in T. *thermophilus* (Gregory *et al.*, 2005a).
30. A2450G-C2063U: exploration of the conserved A + C wobble pair within the ribosomal peptidyl transferase center using purified mutant ribosomes (Hesslein *et al.*, 2004).
31. A2450G-C2063U: resistance mutations in 23S rRNA identify the site of action of the protein synthesis inhibitor linezolid in the ribosomal peptidyl transferase center (Kloss *et al.*, 1999).
32. C2063U, A2450G-C2063U: the A2453-C2499 wobble base pair is responsible for pH sensitivity of the peptidyl transferase active site conformation (Bayfield *et al.*, 2004).
33. G2252U stimulates mRNA slippage in the ribosome. This effect is suppressed by C2394A (McGarry *et al.*, 2005).
34. G2252C inhibits sparsomycin-dependent translocation. The effects of this P-loop mutation are suppressed by pairing with compensatory base change, C74G, in A-site tRNA, but not by pairing with the same compensatory change in P-site tRNAs (Dorner *et al.*, 2006).

35. U2252C-G2253C-G2254A: mutations in B. subtilis that inactivate its protein-folding property in vitro (Chowdhury et al., 2002).
36. C2394A inhibits P/E hybrid state formation in the ribosome. This mutation also suppresses the effects of G2252U (McGarry et al., 2005).
37. C2394G destabilizes the E-site binding in the ribosome and causes increased frameshifting and stop codon readthrough. This mutation affects formation of the P/E hybrid site (Sergiev et al., 2005).
38. G2447A: catalytic properties of mutant 23S ribosomes resistant to oxazolidinones (Bobkova et al., 2003).
39. G2447A confers chloramphenicol resistance in T. thermophilus (Gregory et al., 2005a).
40. G2447 mutations have very little effect on the rate of peptide bond formation (Beringer et al., 2003).
41. G2447: ribosomal peptidyl transferase can withstand mutations at the putative catalytic nucleotide (Polacek et al., 2001).
42. G2447U, ribosomal and nonribosomal resistance to oxalolidinones, species-specific idiosyncrasy of ribosomal alterations (Sander et al., 2002).
43. G2447: mutations in the peptidyl transferase active site (Thompson et al., 2001).
44. G2447U confers oxazolidinone resistance (Xiong et al., 2000).
45. G2447U: pK_a of adenine 2451 in the ribosomal peptidyl transferase center remains elusive (Xiong et al., 2001).
46. A2450G, A2450G-C2063U: the A2453-C2499 wobble base pair is responsible for pH sensitivity of the peptidyl transferase active site conformation (Bayfield et al., 2004).
47. A2450G-C2063U: exploration of the conserved A + C wobble pair within the ribosomal peptidyl transferase center using purified mutant ribosomes (Hesslein et al., 2004).
48. A2450G, A2453G, C2499U, A2450G-C2063U, A2453-C2449U: the A2453-C2499 wobble base pair is responsible for pH sensitivity of the peptidyl transferase active site conformation (Bayfield et al., 2004).
49. A2450G-C2063U: resistance mutations in 23S rRNA identify the site of action of the protein synthesis inhibitor linezolid in the ribosomal peptidyl transferase center (Kloss et al., 1999).
50. A2450G-C2063U: exploration of the conserved A + C wobble pair within the ribosomal peptidyl transferase center using purified mutant ribosomes (Hesslein et al., 2004).
51. A2451 mutations have moderate to no effect on the rate of peptide bond formation depending on the source of ribosomes and the nature of the A-site substrate (Youngman et al., 2004).
52. Deletion or modification of 2451: chemical engineering reveals an important role of the 2′-hydroxyl group of A2451 (Erlacher et al., 2005).

53. A2451: ribosomal peptidyl transferase can withstand mutations at the putative catalytic nucleotide (Gregory et al., 2001a; Polacek et al., 2001).
54. A2451U does not abolish the pH dependence of peptide bond formation in the presence of puromycin in M. smegmatis. The mutation alters the structure of the peptidyl transferase center and changes the pattern of pH-dependent rearrangments. The effects of A2451U are different in E. coli and M. smegmatis; in E. coli, any mutation of A2451 is lethal (Beringer et al., 2005).
55. A2451: mutations in the peptidyl transferase active site (Thompson et al., 2001).
56. A2451U, G, or C is deficient in peptidyl transfer with minimal aminoacyl substrate puromycin, but not with intact aminoacyl-tRNA substrates (Brunelle et al., 2006).
57. C2452U confers chloramphenicol and sparsomycin resistance in T. thermophilus (Gregory et al., 2005a).
58. A2453G confers chloramphenicol resistance in T. thermophilus (Gregory et al., 2005a).
59. A2453G, A2453-C2449U: the A2453-C2499 wobble base pair is responsible for pH sensitivity of the peptidyl transferase active site conformation (Bayfield et al., 2004).
60. A2469C, C2480U: evernimicin (SCH27899) inhibits a novel ribosome target site (Adrian et al., 2000).
61. G2495A mutation confers chloramphenicol resistance in mitochondria of Chinese hamster cells (Hashiguchi and Ikushima, 2000).
62. C2499U, A2453-C2449U: the A2453-C2499 wobble base pair is responsible for pH sensitivity of the peptidyl transferase active site conformation (Bayfield et al., 2004).
63. U2500C confers chloramphenicol resistance in T. thermophilus (Gregory et al., 2005a).
64. A2503G confers chloramphenicol resistance in T. thermophilus (Gregory et al., 2005a).
65. U2504C, G, or A confers chloramphenicol resistance in T. thermophilus (Gregory et al., 2005a).
66. U2504C-U2505A-U2506C: a mutation in domain V of the 23S rRNA of B. subtilis that inactivates its protein-folding property in vitro (Chowdhury et al., 2002).
67. G2505A confers chloramphenicol resistance in T. thermophilus (Gregory et al., 2005a).
68. U2506 mutants exhibit reduced peptidyl transfer with puromycin and mild reduction in peptide release (Youngman et al., 2004).
69. U2506A, G, or C is deficient in peptidyl transfer with minimal aminoacyl substrate puromycin, but not with intact aminoacyl-tRNA substrates (Brunelle et al., 2006).

70. A2531G, a lethal mutation, suppresses the lethal mutation A2662C (Chan et al., 2006).

71. G2535A, G2536C confer resistance to evernimicin (SCH27899) (Adrian et al., 2000).

72. U2548C-U2585C: a mutation of B. subtilis that inactivates its protein-folding property in vitro (Chowdhury et al., 2002).

73. G2551C inhibits sparsomycin-dependent translocation. The effects of this P-loop mutation are suppressed by pairing with compensatory base change, C75G, in A-site tRNA, but not by pairing with the same compensatory change in P-site tRNAs (Dorner et al., 2006).

74. G2553A, C, or U mutations increase the rate of translocation relative to wild type. Pairing these A-loop mutations with compensatory C75G changes in A-site tRNA results in diminished translocation (Dorner et al., 2006).

75. U2585C, G, or A is deficient in peptidyl transfer with minimal aminoacyl substrate puromycin, but not with intact aminoacyl-tRNA substrates (Brunelle et al., 2006).

76. U2585 mutants exhibit reduced peptidyl transfer with puromycin and reduction in RF1-mediated peptide release (Youngman et al., 2004).

77. A2602 mutants exhibit reduced peptidyl transfer with puromycin and mild reduction in peptide release (Youngman et al., 2004).

78. A2602G-U2609C: mutations in B. subtilis that inactivate its protein-folding property in vitro (Chowdhury et al., 2002).

79. A2602C exhibits reduced RF1-mediated peptide release (Youngman et al., 2004).

80. A2602C: mutation confers resistance to 16-membered macrolides and streptogramins in S. pneumoniae (Depardieu and Courvalin, 2001).

81. A2602C, U, or G is deficient in peptidyl transfer with minimal aminoacyl substrate puromycin, but not with intact aminoacyl-tRNA substrates (Brunelle et al., 2006).

82. U2609C: binding site of the bridged macrolides in the E. coli ribosome (Xiong et al., 2005).

83. U2609C, binding site of macrolide antibiotics on the ribosome, new resistance mutation identifies a specific interaction of deltoids with rRNA (Garza-Ramos et al., 2001).

84. U2609C eliminates tryptophan induction of tna operon expression (Cruz-Vera et al., 2005).

D. Mutations in domain VI of 23S rRNA

1. G2655C, G2655A, G2655U: affinity purification of ribosomes with a lethal G2655C mutation that affects translocation (Leonov et al., 2003).

2. G2655C or G2655U abolish binding of EF-G (Chan et al., 2006).

3. A2660, G2661: mutations at these nucleotides do not affect EF-G binding (Chan et al., 2006).
4. A2662C, a lethal mutation, is suppressed by a lethal mutation at A2531 (Chan et al., 2006).

IV. MUTATIONAL ANALYSIS OF RIBOSOMAL PROTEIN STRUCTURE AND FUNCTION

Ribosomal proteins are named according to subunit (Brodersen and Nissen, 2005; Liljas, 2004). Examples of the results of mutational analysis of ribosomal protein structure and function are organized accordingly.

A. Mutations in small-subunit proteins

1. S1, absent (Dabbs et al., 1981); S1, S9, and S20 absent (Dabbs et al., 1983).
2. S1 mutations affect recognition of TmRNA (Okada et al., 2004).
3. 131, 132, 135 in S3: mutations affect mRNA helicase activity in the ribosome (Takyer et al., 2005).
4. S4 mutations influence the higher order structure of 16S rRNA (Allen and Noller, 1989).
5. S4: mutants affect regulation of ribosome accumulation in E. coli (Mikkola and Kurland, 1991).
6. S4: mutants with altered accuracy (Dong and Kurland, 1995).
7. S4, 535, frameshift; S12, 87; double mutant: a mutator phenotype that correlates with increased mistranslation (Balashov and Jumayun, 2003).
8. 44, 47 in S4: mutations affect mRNA helicase activity in the ribosome (Takyer et al., 2005).
9. 28 in S5 affects nonsense readthrough in B. subtilis (Inaoka et al., 2001).
10. 19, 28 in S5: mutation affects mRNA helicase activity in the ribosome (Takyer et al., 2005).
11. 2, 3, 9, 14, deletion 1–17, 34, 35, 34 and 35, 52, 73, 77 and 78, 84, 85, 86, 89, deletion 73–89, 94, 101, 102, 110, 113, 118, 122, 125, 135, 136, 137, 138, 139, 142, 143, 145, 147, 148, 135 and 139, 142, deletion 130–178, deletion 409, deletion 403 in S7: mutants defective in assembly and function of 30S subunits (Fredrick et al., 2000).
12. Deletion of 148–178, 156–178 in S7; substitution of 148–155 in S7: defines a functional interaction between S7 and S11 (Robert and Brakier-Gingras, 2003).
13. S8 mutations (reviewed in Zimmermann et al., 2000).
14. S9 absent; S1, S9, and S20 absent (Dabbs et al., 1983).

15. Deletion of C-terminal tails from S9 and S13 creates ribosomes with all-RNA 30S subunit (Hoang *et al.*, 2004; Noller *et al.*, 2005).
16. S11 and S12, S11 and S12 and S13 proteins absent. Cysteine substitutions 70 and 121 in S11. Proteins S12 and S13 function as control elements for translocation of the mRNA:tRNA complex (Cukras *et al.*, 2003).
17. Substitution of 55–63 in S11: defines a functional interaction between S7 and S11 (Robert and Brakier-Gingras, 2003).
18. S12 mutants express processivity errors in *E. coli* (Jorgensen and Kurland, 1990).
19. S12 mutants affect regulation of ribosome accumulation in *E. coli* (Mikkola and Kurland, 1991).
20. S12: kinetic properties of mutant *E. coli* ribosomes (Bilgin *et al.*, 1992).
21. S12 mutations influence the higher order structure of 16S rRNA (Allen and Noller, 1989).
22. S12: mutants with altered accuracy (Dong and Kurland, 1995).
23. 56 in S12 affects nonsense readthrough in *B. subtilis* (Inaoka *et al.*, 2001).
24. S12, S12 and S13, S11 and S12, S11 and S12 and S13 proteins absent. Cysteine substitutions 104, 27, 34, 53 in S12. Proteins S12 and S13 function as control elements for translocation of the mRNA:tRNA complex (Cukras *et al.*, 2003).
25. 90, 37, 41, double mutants, 103, 201 in S12: streptomycin-resistant and streptomycin-dependent mutants in *T. thermophilus* (Carr *et al.*, 2005).
26. 41, 42, 85, 87, 90 in S12: streptomycin resistance in *T. thermophilus* IB-21 (Gregory *et al.*, 2001b).
27. S12, 87; S4, 535, frameshift; double mutant: a mutator phenotype that correlates with increased mistranslation (Balashov and Jumayun, 2003).
28. 88 in S12 confers streptomycin resistance in *B. burgdorferi* (Criswell *et al.*, 2006).
29. Cysteine substitutions 87 in S13. Proteins S12 and S13 function as control elements for translocation of the mRNA:tRNA complex (Cukras *et al.*, 2003).
30. Deletion of C-terminal tails from S9 and S13 creates ribosomes with all-RNA 30S subunit (Hoang *et al.*, 2004; Noller *et al.*, 2005).
31. S15: pseudoknot structure of S15 translational operator from *E. coli* (Benard *et al.*, 1994).
32. S15 deletions reveal RNA structural adaptation permitting S15 repression of translation of its own mRNA (Serganov *et al.*, 2003).
33. Both S17 and L29 absent (Dabbs *et al.*, 1983).
34. S20 absent. S1, S9, and S20 absent (Dabbs *et al.*, 1983).
35. S20 and L11 deletion: subunit association defects (Gotz *et al.*, 1989).
36. Cysteine substitutions 23 in S21 (Cukras *et al.*, 2003).

B. Mutations in large-subunit proteins

1. L1 absent (Dabbs et al., 1981).
2. L1, L2, L15, L19, L24, L27, L29, L30, L33 each absent. L28 and L33 both absent. S17 and L29 both absent (Dabbs et al., 1983).
3. 68, 70, 86, 155, 66, 95, 102 in L2: 23S rRNA-binding site (Harada et al., 1998).
4. L2: mutational analysis (reviewed in Tanaka et al., 2000).
5. 149, 445 in L3 confer resistance to tiamulin (Bosling et al., 2003).
6. L3: saturation mutagenesis results (reviewed in Meskauskas et al., 2005).
7. L4 mutations confer erythromycin resistance (Gregory and Dahlberg, 1999).
8. L4 mutations confer ketolide and azalide resistance (Schlunzen et al., 2003).
9. Substitution 69, 70, 71, insertion at 69 in L4: mutations confer macrolide resistance in S. pneumoniae (Tait-Kamradt et al., 2000a).
10. Substitution 63–74, 69, insertion at 67 and 68 in L4: macrolide resistance conferred in pneumococcal strains (Tait-Kamradt et al., 2000b).
11. 63 in L4: erythromycin-resistant mutants (Gabashvili et al., 2001).
12. 71 in L4: mutations confer resistance to macrolides, clindamycin, streptogramin, and telithromycin in S. pneumoniae (Canu et al., 2002).
13. Insertion at 69 in L4 confers resistance to macrolides in S. pyogenes (Malbruny et al., 2002b).
14. L4: macrolide-binding site (Hansen et al., 2004).
15. L4: mutations confer resistance to erythromycin in T. thermophilus (Tsagkalia et al., 2005).
16. 70 in L4, insertions at 60 in L4 confer resistance to macrolides in M. pneumoniae (Pereyre et al., 2004).
17. 63 in L4: multiple effects on translation (O'Connor et al., 2004).
18. Glu82Lys in L7/L12: mutation affects growth and translational accuracy (Kirsebom et al., 1986).
19. Deletion of 38–46, 44–52, or 38–52 in L7/L12 (Gudkov et al., 1991).
20. L10, deletion of 10, 20, or 33 amino acids from C-terminus of L10 affects L7/L12 binding (Griaznova and Traut, 2000).
21. L11: mutational analysis of RNA recognition and binding (reviewed in Draper et al., 2000).
22. L11 inducible knockout, C-terminal amino acids 68–142 only: mutations affect interaction of thiostrepton and EF-G with L11-binding domain (Bowen et al., 2005).
23. L11 and S20 deletion affects subunit association in E. coli ribosomes (Gotz et al., 1989).
24. 21, deletion of 20–21, 35, 55, 22 in L11: mutations confers resistance to thiostrepton in T. thermophilus (Cameron et al., 2004).

25. L16: mutations confer evernimicin resistance (Belova *et al.*, 2001).
26. L20, deletion of N-terminal tail in L20 affects ribosome assembly (Guillier *et al.*, 2005).
27. L22: mutations confer resistance to streptogramins (Canu and Leclercq, 2001).
28. L22, 95, 99, 93, 91, 83, 93 and 91 and 83 in L22: mutations confer resistance to macrolides, clindamycin, streptogramin, and telithromycin in *S. pneumoniae* (Canu *et al.*, 2002).
29. K90 substitutions in L22 reduce or eliminate tryptophan induction of tna operon expression (Cruz-Vera *et al.*, 2005).
30. L22, 82 and 83 and 84 triple amino acid alteration in L22 confers erythromycin resistance (Davydova *et al.*, 2002).
31. Deletion 82, 83, and 84 in L22: erythromycin-resistant mutants (Gabashvilli *et al.*, 2001).
32. L22 mutations confer erythromycin resistance (Gregory and Dahlberg, 1999).
33. L22: macrolide-binding site (Hansen *et al.*, 2004).
34. L22: mutations confer resistance to macrolides in *S. pyogenes* (Malbruny *et al.*, 2002b).
35. C-terminal tail insertions and deletions in L22 confer quinupristin-dalfopristin resistance in *S. aureus* (Malbruny *et al.*, 2002a).
36. Deletion of 82–84 in L22: multiple effects on translation (O'Connor *et al.*, 2004).
37. L22, deletion of 101–103 in L22 confers antibiotic resistance (Prunier *et al.*, 2002).
38. L22 mutations confer ketolide and azalide resistance (Schlunzen *et al.*, 2003).
39. 90 in L22: mutations reduce or abolish tryptophan induction of tna operon expression (Cruz-Vera *et al.*, 2005).
40. 94 in L22 confers telithromycin resistance in laboratory-generated mutants of *S. pneumoniae* (Walsh *et al.*, 2003).
41. L27 deletion results in severe growth defects; deletion of three N-terminal amino acids in L27 leads to a decrease in growth rate and impairment in peptidyl transferase activity (Maguire *et al.*, 2005 and references therein).
42. Both L29 and S17 absent (Dabbs *et al.*, 1983).
43. L30 in yeast (a member of the L7/L12 family of proteins): internal loop mutations affect the binding of L30 to its own message (White *et al.*, 2004).

V. MUTATIONAL ANALYSIS OF RIBOSOMAL FACTOR STRUCTURE AND FUNCTION

A. Mutations in initiation factors

The details of the roles of initiation factors (IFs) in translation have been reviewed elsewhere (Gualerzi *et al.*, 2000; Laursen *et al.*, 2005; Roll-Mecak *et al.*, 2001).

1. 345, 346, 347, 384, 385, 386 in *in vitro* and *in vivo* eukaryotic IF: subunit interactions between eIF2 and eIF5 studied in mammalian system (Das and Maitra, 2000).
2. Thr445, Ile500 in IF2: mutations in G-domain confer cold-sensitive growth in *E. coli* (Larigauderie *et al.*, 2000).

B. Mutations in elongation factors

The details of the roles of elongation factors (EFs) in protein synthesis have been reviewed elsewhere (Andersen *et al.*, 2003; Hilgenfeld *et al.*, 2000; Rodnina *et al.*, 2000, 2005; Spahn *et al.*, 2004).

1. Mutations in EFTu
   a. 120, 124, 160, 298, 316, 329, 375, 378 in EFTu: mutations confer kirromycin resistance (Abdulkarim *et al.*, 1994).
   b. 204 in EFTu (Andersen *et al.*, 2003).
   c. EFTu mutants in *T. thermophilus* (Zeidler *et al.*, 1995).
   d. His84Arg in EFTu: streptomycin interferes with coupling of codon recognition and GTPase activation (Gromadski and Rodnina, 2004).
2. Mutations in EF-G
   a. Deletion G domain in EF-G promotes translocation of the 3′ end but not of the anticodon domain of peptidyl-tRNA (Borowski *et al.*, 1996).
   b. 16, 84 in EF-G: fusidic acid resistance (Hansson *et al.*, 2005).
   c. 495, 502, 563, 608, deletion 608–703 in EF-G: C-terminal amino acid residues essential for ribosome association and translocation (Hou *et al.*, 1994b).
   d. 495, 502 in EF-G: conditional-lethal mutations (Hou *et al.*, 1994a).
   e. 583, deletion of domain 1, 4, or 5 in EF-G: release of ribosome-bound ribosome-recycling factor (RRF) (Kiel *et al.*, 2003).
   f. 504, 554, 534, insertion of 501–504 in EF-G; functional role of loop region of EF-G in translocation (Kolesnikov and Gudkov, 2003).
   g. 502 in EF-G: temperature-sensitive mutation suppressed either by C1192U in 16S rRNA and G2058A in 23S rRNA or by C1400U in 16S rRNA and G2058A in 23S rRNA (Koosha *et al.*, 2000).
   h. Fusidic acid-resistant mutants of EF-G (Liljas *et al.*, 2000).
   i. Fusidic acid-resistant EF-G mutants of *Salmonella enterica serovar typhimurium* (Macvanin *et al.*, 2003).
   j. 583 in EF-G, double mutant bridge between domains 1–5 induced structural change in helix 34 of 16S rRNA related to translocation (Matassova *et al.*, 2001).
   k. 29, 59 in EF-G: arginines important for GTP hydrolysis or translocation (Mohr *et al.*, 2000).
   l. 30, 32, 37, 38, 41 in EF-G: conserved amino acid residues adjacent to the effector domain, substitution S7 28 amino acids (Sharer *et al.*, 1999).

C. Mutations in release factors

The details of the roles of RFs in termination of protein synthesis have been reviewed elsewhere (Nakamura et al., 2000; Wilson et al., 2000).

a. 181, 182, 183, 184 in eukaryotic RF1 abolish ability of human eRF1 to trigger peptidyl-tRNA hydrolysis (Frolova et al., 1999).
b. Transposon insertions in RF3 and RF4 demonstrate role of factors in dissociation of peptidyl-tRNA from the ribosome (Heurgue-Hamard et al., 1998).
c. RF1, RF2 GG to GGA, GA, chimeras in RF1 and RF2 define the role of RFs in termination (Kisselev et al., 2002).
d. RF1: temperature sensitivity caused by mutant RF1 is suppressed by mutations that affect 16S rRNA maturation (Kaczanowska and Ryden-Aulin, 2004).
e. Arg173Pro in RF1: an interaction between RF1 and ribosomal protein L7/L12 in vivo (Zhang et al., 1994).

D. Mutations in ribosome recycling factor

The details of the role of RRF in protein synthesis and the results of mutational analysis of RRF have been reviewed elsewhere (Kaji and Hirokawa, 2000).

VI. CONCLUDING REMARKS

Previous reviews tabulated mutations in small-subunit rRNA (Triman, 1995) and in large-subunit rRNA (Triman, 1999). A text-based database has provided online access to the tabulations since 1994; a more convenient searchable version of the database was subsequently made available at http://ribosome.fandm.edu. Resource limitations have hampered attempts to update the rRNA mutation dataset and to expand the scope of the database to include mutations in ribosomal proteins and ribosomal factors. The resource issue is one that affects the majority of existing databases (Merali and Giles, 2005). Ideally, the summary of examples of relevant literature included in this chapter will provide a template for expansion of the rRNA mutation database to a broader, more useful online resource that includes mutations in ribosomal proteins, factors, and, ultimately, tRNAs.

Two annual issues of the journal, Nucleic Acids Research, the Database issue and the Web Server issue, update and maintain a collection of useful online resources (Bateman, 2006; Galperin, 2006). The compilation of links to molecular biology servers in The Bioinformatics Links Directory promises to facilitate efficient access to these resources (Fox et al., 2006). The RNA World

Web site maintained by Juergen Suhnel continues to play an important role as a catalog of relevant sites for the RNA community (Hammann, 2003). A promising approach to solving the problem of providing centralized access to RNA data is represented by the initiatives and efforts of the RNA Ontology Consortium (Leontis *et al.*, 2006).

Acknowledgments

This work was supported by grants from the National Science Foundation and the Provost of Franklin and Marshall College. K.L.T. is particularly grateful to Steven Vavoulis for his help with the preparation of this chapter and to Janan Eppig, Lea Brakier-Gingras, Albert Dahlberg, Steven Gregory, Alexander Mankin, Harry Noller, Michael O'Connor, and Catherine Squires for their support and encouragement.

References

Abdi, N. M., and Fredrick, K. (2005). Contribution of 16S rRNA nucleotides forming the 30S subunit A and P sites to translation in *Escherichia coli*. *RNA* **11,** 1624–1632.

Abdulkarim, F., Liljas, A., and Hughes, D. (1994). Mutations to kirromycin resistance occur in the interface of domains I and III of EF-Tu. GTP. *FEBS Lett.* **352,** 118–122.

Adrian, P. V., Mendrick, C., Loebenberg, D., McNicholas, P., Shaw, K. J., Klugman, K. P., Hare, R. S., and Black, T. A. (2000). Evernimicin (SCH27899) inhibits a novel ribosome target site: Analysis of 23S ribosomal DNA mutants. *Antimicrob. Agents Chemother.* **44,** 3101–3106.

Allen, P. N., and Noller, H. F. (1989). Mutations in ribosomal proteins S4 and S12 influence the higher order structure of 16S ribosomal RNA. *J. Mol. Biol.* **208,** 457–468.

Andersen, G. R., Nissen, P., and Nyborg, J. (2003). Elongation factors in protein biosynthesis. *Trends Biochem. Sci.* **28,** 434–441.

Arkov, A. L., Freistroffer, D. V., Ehrenberg, M., and Murgola, E. J. (1998). Mutations in RNAs of both ribosomal subunits cause defects in translation termination. *EMBO J.* **17,** 1507–1514.

Arkov, A. L., Hedenstierna, K. O. F., and Murgola, E. J. (2002). Mutational evidence for a functional connection between two domains of 23SrRNA in translation termination. *J. Bacteriol.* **184,** 5052–5057.

Balashov, S., and Jumayun, M. Z. (2003). *Eschericia coli* cells bearing a ribosomal ambiguity mutation in rpsD have a mutator phenotype that correlates with increased mistranslation. *J. Bacteriol.* **185,** 5015–5018.

Bateman, A. (2006). Editorial. *Nucleic Acids Res.* **34,** (Database issue), D1.

Bayfield, M. A., Thompson, J., and Dahlberg, A. E. (2004). The A2453-C2499 wobble base pair is responsible for pH sensitivity of the peptidyltransferase active site conformation. *Nucleic Acids Res.* **32,** 5512–5518.

Belanger, F., Leger, M., Sariya, A. A., Cunningham, P. R., and Brakier-Gingras, L. (2002). Functional studies of the 900 tetraloop capping helix 27 of 16S ribosomal RNA. *J. Mol. Biol.* **320,** 979–989.

Belanger, F., Theberge-Julien, G., Cunningham, P. R., and Brakier-Gingras, L. (2005). A functional relationship between helix I and the 900 tetraloop of 16S ribosomal RNA within the bacterial ribosome. *RNA* **11,** 903–913.

Belova, L., Tenson, T., Xiong, L., McNicholas, P. M., and Mankin, A. S. (2001). A novel site of antibiotic action in the ribosome: Interaction of evernimicin with the large ribosomal subunit. *Proc. Natl. Acad. Sci. USA* **98**, 3726–3731.

Benard, L., Philippe, C., Dondon, L., Grunberg-Manago, M., Ehresmann, B., Ehresmann, C., and Portier, C. (1994). Mutational analysis of the pseudoknot structure of the S15 translational operator from *Escherichia coli*. *Mol. Microbiol.* **14**, 31–40.

Beringer, M., Adio, S., Wintermeyer, W., and Rodnina, M. (2003). The G2447A mutation does not affect ionization of a ribosomal group taking part in peptide bond formation. *RNA* **9**, 919–922.

Beringer, M., Bruell, C., Xiong, L., Pfister, P., Bieling, P., Katunin, V. I., Mankin, A. S., Böttger, E. C., and Rodnina, M. V. (2005). Essential mechanisms in the catalysis of peptide bond formation on the ribosome. *J. Biol. Chem.* **280**, 36065–36072.

Bilgin, N., Claesens, F., Pahverk, H., and Ehrenberg, M. (1992). Kinetic properties of *Escherichia coli* ribosomes with altered forms of S12. *J. Mol. Biol.* **224**, 1011–1027.

Bobkova, E. V., Yan, Y. P., Jordan, D. B., Kurilla, M. G., and Pompliano, D. L. (2003). Catalytic properties of mutant 23S ribosomes resistant to oxazolidinones. *J. Biol. Chem.* **278**, 9802–9807.

Borowski, C., Rodnina, M. V., and Wintermeyer, W. (1996). Truncated elongation factor G lacking the G domain promotes translocation of the 3′ end but not of the anticodon domain of peptidyl-tRNA. *Proc. Natl. Acad. Sci. USA* **93**, 4202–4206.

Bosling, J., Poulsen, S. M., Vester, B., and Long, K. S. (2003). Resistance to the peptidyl transferase inhibitor tiamulin caused by a mutation of ribosomal protein L3. *Antimicrob. Agents Chemother.* **47**, 2892–2896.

Bowen, W. S., VanDyke, N., Murgola, E. J., Lodmell, J. S., and Hill, W. E. (2005). Interaction of thiostrepton and elongation factor-G with the ribosomal protein L11-binding domain. *J. Biol. Chem.* **280**, 2934–2943.

Brodersen, D. E., and Nissen, P. (2005). The social life of ribosomal proteins. *FEBS J.* **272**, 2098–2108.

Brunelle, J. L., Youngman, E. M., Sharma, D., and Green, R. (2006). The interaction between C75 of tRNA and the A loop of the ribosome stimulates peptidyl transferase activity. *RNA* **12**, 33–39.

Brunelli, C. A., O'Connor, M., and Dahlberg, A. E. (2002). Decreased requirement for 4.5S RNA in 16S and 23S rRNA mutants of *Escherichia coli*. *FEBS Lett.* **514**, 44–48.

Cameron, D. M., Thompson, J., Gregory, S. T., March, P. E., and Dahlberg, A. E. (2004). Thiostrepton-resistant mutants of *Thermus thermophilus*. *Nucleic Acids Res.* **32**, 3220–3227.

Camps, M., Arrizabalanga, G., and Boothroyd, J. (2002). A rRNA mutation identifies the apicoplast as the target for clindamycin in *Toxoplasma gondii*. *Mol. Microbiol.* **43**, 1309–1318.

Cannone, J. J., Subramanian, S., Schnare, M. N., Collett, J. R., D'Souza, L. M., Du, Y., Feng, B., Lin, N., Madabusi, L. V., Müller, K. M., Pande, N., Shang, Z., *et al.* (2002). The comparative RNA web (CRW) site: An online database of comparative sequence and structure information for ribosomal, intron, and other RNAs. *BMC Bioinformatics* **3**, 2.

Canu, A., and Leclercq, R. (2001). Overcoming bacterial resistance by dual target inhibition: The case of streptogramins. *Curr. Drug Targets Infect. Disord.* **1**, 215–225.

Canu, A., Malbruny, B., Coquemont, M., Davies, T. A., Appelbaun, P. C., and Leclercq, R. (2002). Diversity of ribosomal mutations conferring resistance to macrolides, clindamycin, streptogramin, and telithromycin in *Streptococcus pneumoniae*. *Antimicrob. Agents Chemother.* **46**, 125–131.

Carr, J. F., Gregory, S. T., and Dahlberg, A. E. (2005). Severity of the streptomycin resistance and streptomycin dependence phenotypes of ribosomal protein S12 of *Thermus thermophilus* depends on the identity of highly conserved amino acid residues. *J. Bacteriol.* **187**, 3548–3550.

Chan, Y.-L., Dresios, J., and Wool, I. G. (2006). A pathway for the transmission of allosteric signals in the ribosome through a network of RNA tertiary interactions. *J. Mol. Biol.* **355**, 1014–1025.

Chernyaeva, N. S., and Murgola, E. J. (2000). Covariance of complementary rRNA loop nucleotides does not necessarily represent functional pseudoknot formation *in vivo. J. Bacteriol.* **182**, 5671–5675.

Chowdhury, S., Pal, S., Ghosh, J., and DasGupta, C. (2002). Mutations in domain V of the 23S ribosomal RNA of *Bacillus subtilis* that inactivate its protein folding property *in vitro. Nucleic Acids Res.* **30**, 1278–1285.

Cochella, L., and Green, R. (2004). Isolation of antibiotic resistance mutations in the rRNA by using an *in vitro* selection system. *Proc. Natl. Acad. Sci. USA* **101**, 3786–3791.

Cochella, L., and Green, R. (2005). An active role for tRNA in decoding beyond codon:anticodon pairing. *Science* **308**, 1178–1180.

Criswell, D., Tobiason, V. L., Lodmell, J. S., and Samuels, D. S. (2006). Mutations conferring aminoglycoside and spectinomycin resistance in *Borrelia burgdorferi. Antimicrob. Agents Chemother.* **50**, 445–452.

Cruz-Vera, L. R., Rajagopal, S., Squires, C., and Yanofsky, C. (2005). Features of ribosome-peptidyl-tRNA interactions essential for tryptophan induction of *tna* operon expression. *Mol. Cell* **19**, 333–343.

Cukras, A. R., Southworth, D. R., Brunelle, J. L., Culver, G. M., and Green, R. (2003). Ribosomal proteins S12 and S13 function as control elements for translocation of the mRNA:tRNA complex. *Mol. Cell* **12**, 321–328.

Dabbs, E. R., Ehrlich, R., Hasenbank, R., Schroeter, B.-H., Stoffler-Meilicke, M., and Stoffler, G. (1981). Mutants of *Escherichia coli* lacking ribosomal protein L1. *J. Mol. Biol.* **149**, 553–578.

Dabbs, E. R., Hasenbank, R., Kastner, B., Rak, K.-H., Wartusch, B., and Stoffler, G. (1983). Immunological studies of *Escherichia coli* mutants lacking one or two ribosomal proteins. *Mol. Gen. Genet.* **192**, 301–308.

Dailidiene, D., Bertoli, M. T., Miciuleviciene, J., Mukhopadhyay, A. K., Dailide, G., Pascasio, M. A., Kupcinskas, L., and Berg, D. E. (2002). Emergence of tetracycline resistance in *Helicobacter pylori*: Multiple mutational changes in 16S ribosomal DNA and other genetic loci. *Antimicrob. Agents Chemother.* **46**, 3940–3946.

Das, S., and Maitra, U. (2000). Mutational analysis of mammalian translation initiation factor 5 (eIF5): Role of interaction between the beta subunit of eIF2 and eIF5 in eIF5 function *in vitro* and *in vivo. Mol. Cell. Biol.* **20**, 3942–3950.

Davydova, N., Streltsov, V., Wilce, M., Liljas, A., and Garber, M. (2002). L22 ribosomal protein and effect of its mutation on ribosome resistance to erythromycin. *J. Mol. Biol.* **322**, 635–644.

Depardieu, F., and Courvalin, P. (2001). Mutation in 23S rRNA responsible for resistance to 16-membered macrolides and streptogramins in *Streptococcus pneumoniae. Antimicrob. Agents Chemother.* **45**, 319–323.

Dong, H., and Kurland, C. G. (1995). Ribosome mutants with altered accuracy translate with reduced processivity. *J. Mol. Biol.* **248**, 551–561.

Dorner, S., Brunelle, J. L., Sharma, D., and Green, R. (2006). The hybrid state of tRNA binding is an authentic translation elongation intermediate. *Nat. Struct. Mol. Biol.* **13**, 234–241.

Draper, D. E., Conn, G. L., Gittes, A. G., Guhathakurta, D., Lattman, E. E., and Reynaldo, L. (2000). RNA tertiary structure and protein recognition in an L11-RNA complex. *In* "The Ribosome: Structure, Function, Antibiotics and Cellular Interactions" (R. A. Garrett, S. R. Douthwaite, A. Liljas, A. T. Matheson, P. B. Moore, and H. Noller, eds.), pp. 105–114. ASM Press, Washington, DC.

Erlacher, M. D., Lang, K., Shankaran, N., Wotzel, B., Huttenhofer, A., Micura, R., Mankin, A. S., and Polacek, N. (2005). Chemical engineering of the peptidyl transferase center reveals an important role of the 2′-hydroxyl group of A2451. *Nucleic Acids Res.* **33**, 1618–1627.

Fox, J. A., McMillan, S., and Ouellette, B. F. F. (2006). A compilation of molecular biology web servers: 2006 update on the Bioinformatics Links Directory. *Nucleic Acids Res.* **34**(Web Server issue), W3–W5.

Fredrick, K., Dunny, G. M., and Noller, H. F. (2000). Tagging ribosomal protein S7 allows rapid identification of mutants defective in assembly and function of 30S subunits. *J. Mol. Biol.* **298,** 379–394.

Frolova, L. Y., Tsivkovskii, R. Y., Sivolobova, G. F., Oparina, N. Y., Serpinsky, O. I., Blinov, V. M., Tatkov, S. I., and Kisselev, L. L. (1999). Mutations in the highly conserved GG motif of class I polypeptide release factors abolish ability of human eRF1 to trigger peptidyl-tRNA hydrolysis. *RNA* **5,** 1014–1020.

Furneri, P. M., Rappazzo, G., Musumarra, M. P., DiPietro, P., Catania, L. S., and Roccasalva, L. S. (2001). Two new point mutations at A2062 associated with resistance to 16-membered macrolide antibiotics in mutant strains of *Mycoplasma hominis*. *Antimicrob. Agents Chemother.* **45,** 2958–2960.

Gabashvilli, I. S., Gregory, S. T., Valle, M., Grassucci, R., Worbs, M., Wahl, M. C., Dahlberg, A. E., and Frank, J. (2001). The polypeptide tunnel system in the ribosome and its gating in erythromycin resistance mutants of L4 and L22. *Mol. Cell* **8,** 181–188.

Galperin, M. Y. (2006). The Molecular Biology Database Collection: 2006 update. *Nucleic Acids Res.* **34,** Database issue, D3–D5.

Garza-Ramos, G., Xiong, L., Zhong, P., and Mankin, A. (2001). Binding site of macrolike antibiotics on the ribosome: New resistance mutation identifies a specific interaction of ketolides with rRNA. *J. Bacteriol.* **183,** 6898–6907.

Gotz, F., Fleischer, C., Pon, C. L., and Gualerzi, C. O. (1989). Subunit association defects in *Escherichia coli* ribosome mutants lacking proteins S20 and L11. *Eur. J. Biochem.* **183,** 19–24.

Gregory, S. T., and Dahlberg, A. E. (1999). Erythromycin resistance mutations in ribosomal proteins L22 and L4 perturb the higher order structure of 23S ribosomal RNA. *J. Mol. Biol.* **289,** 827–834.

Gregory, S. T., Bayfield, M. A., O'Connor, M., Thompson, J., and Dahlberg, A. E. (2001a). Probing ribosome structure and function by mutagenesis. *Cold Spring Harb. Symp. Quant. Biol.* **66,** 101–108.

Gregory, S. T., Cate, J. H. D., and Dahlberg, A. E. (2001b). Streptomycin-resistant and streptomycin-dependent mutants of the extreme thermophile *Thermus thermophilus*. *J. Mol. Biol.* **309,** 333–338.

Gregory, S. T., Cate, J. H. D., and Dahlberg, A. E. (2001c). Spontaneous erythromycin resistance mutation in a 23S rRNA gene, rrlA, of the extreme thermophile *Thermus thermophilus* IB-21. *J. Bacteriol.* **183,** 4382–4385.

Gregory, S. T., Carr, J. F., and Dahlberg, A. E. (2005a). A mutation in the decoding center of *Thermus thermophilus* 16S rRNA suggest a novel mechanism of streptomycin resistance. *J. Bacteriol.* **187,** 2200–2202

Gregory, S. T., Carr, J. F., Rodriguez-Correa, D., and Dahlberg, A. E. (2005b). Mutational analysis of 16S and 23S genes of. *Thermus thermophilus*. *J. Bacteriol.* **187,** 4804–4812.

Griaznova, O., and Traut, R. R. (2000). Deletion of C-terminal residues of *Escherichia coli* ribosomal protein L10 causes the loss of binding of one L7/L12 dimer: Ribosomes with one L7/L12 dimer are active. *Biochemistry* **39,** 4075–4081.

Gromadski, K. B., and Rodnina, M. V. (2004). Streptomycin interferes with conformational coupling between codon recognition and GTPase activation on the ribosome. *Nat. Struct. Mol. Biol.* **11,** 316–322.

Gualerzi, C. O., Brandi, L., Caserta, E., LaTeana, A., Spurio, R., Tomsic, J., and Pon, C. L. (2000). Translation initiation in bacteria. *In* "The Ribosome: Structure, Function, Antibiotics and Cellular Interactions" (R. A. Garrett, S. R. Douthwaite, A. Liljas, A. T. Matheson, P. B. Moore, and H. Noller, eds.), pp. 477–494. ASM Press, Washington, DC.

Guillier, M., Allemand, F., Graffe, M., Raibaud, M., Dardel, F., Springer, M., and Chiarutini, C. (2005). The N-terminal extension of *Escherichia coli* ribosomal protein L20 is important for ribosome assembly, but dispensable for translational feedback control. *RNA* **11,** 728–738.

Gudkov, A. T., Bubunenko, M. G., and Gryaznova, O. I. (1991). Overexpression of L7/L12 protein with mutations in its flexible region. *Biochimie* **73,** 1387–1389.

Hammann, C. (2003). Web Site: RNA for everyone! The RNA World Website at IMB Jena. *Chembiochem* **4,** 3.

Hansen, J. L., Ippolito, J. A., Ban, N., Nissen, P., Moore, P. B., and Steitz, T. A. (2004). The structure of four macrolide antibiotics bound to the large ribosomal subunit. *Mol. Cell* **10,** 117–128.

Hansson, S., Singh, R., Gudkov, A. T., Liljas, A., and Logan, D. T. (2005). Structural insights into fusidic acid resistance and sensitivity in EF-G. *J. Mol. Biol.* **348,** 939–949.

Harada, N., Maemura, K., Yamasaki, N., and Kimura, M. (1998). Identification by site-directed mutagenesis of amino acid residues in ribosomal protein L2 that are essential for binding to 23S ribosomal RNA. *Biochim. Biophys. Acta* **1429,** 176–186.

Hashiguchi, K., and Ikushima, T. (2000). Novel point mutations in mitochondrial 16S rRNA gene of Chinese hamster cells. *Genes Genet. Syst.* **75,** 59–67.

Hesslein, A. E., Katunin, V. I., Beringer, M., Kosek, A. B., Rodnina, M. V., and Strobel, S. A. (2004). Exploration of the conserved A + C wobble pair within the ribosomal peptidyl transferase center using purified mutant ribosomes. *Nucleic Acids Res.* **32,** 3760–3770.

Heurgue-Hamard, V., Karimi, R., Mora, L., MacDougall, J., Leboeuf, C., Grentzmann, G., Ehrenberg, M., and Buckingham, R. H. (1998). Ribosome release factor RF4 and termination factor RF3 are involved in dissociation of peptidyl-tRNA from the ribosome. *EMBO J.* **17,** 808–816.

Hilgenfeld, R., Mesters, J., and Hoff, T. (2000). Insights into the GTPase mechanism of EF-Tu from structural studies. *In* "The Ribosome: Structure, Function, Antibiotics and Cellular Interactions" (R. A. Garrett, S. R. Douthwaite, A. Liljas, A. T. Matheson, P. B. Moore, and H. Noller, eds.), pp. 347–357. ASM Press, Washington, DC.

Hirabayashi, N., Sato, N. S., and Suzuki, T. (2006). Conserved loop sequence of helix 69 in *Escherichia coli* 23S rRNA is involved in A-site tRNA binding and translational fidelity. *J. Biol. Chem.* **281,** 17203–17211.

Hoang, L., Fredrick, K., and Noller, H. F. (2004). Creating ribosomes with an all-RNA 30S subunit P site. *Proc. Natl. Acad. Sci. USA* **101,** 12439–12443.

Hou, Y., Lin, Y.-P., Sharer, J. D., and March, P. E. (1994a). *In vivo* selection of conditional-lethal mutations in the gene encoding elongation factor G of *Escherichia coli. J. Bacteriol.* **176,** 123–129.

Hou, Y., Yaskowiak, E. S., and March, P. E. (1994b). Carboxy-terminal amino acid residues in elongation factor G essential for ribosome association and translocation. *J. Bacteriol.* **176,** 7038–7044.

Inaoka, T., Kasai, K., and Ochi, K. (2001). Construction of an *in vivo* nonsense read through assay system and functional analysis of ribosomal proteins S12, S4 and S5 in *Bacillus subtilis. J. Bacteriol.* **183,** 4958–4963.

Jamal, M. A., Maeda, S., Nakata, N., Kai, M., Fakuchi, K., and Kashiwabara, Y. (2000). Molecular basis of clarithromycin-resistance in *Mycobacterium avium intracellulare complex. Tuber. Lung Dis.* **80,** 1–4.

Jorgensen, F., and Kurland, C. G. (1990). Processivity errors of gene expression in *Escherichia coli. J. Mol. Biol.* **215,** 51–521.

Kaczanowska, M., and Ryden-Aulin, M. (2004). Temperature sensitivity caused by mutant release factor 1 is suppressed by mutations that affect 16S rRNA maturation. *J. Bacteriol.* **186,** 3046–3055.

Kaji, A., and Hirokawa, G. (2000). Ribosome-recycling factor: An essential factor for protein synthesis. *In* "The Ribosome: Structure, Function, Antibiotics and Cellular Interactions" (R. A. Garrett, S. R. Douthwaite, A. Liljas, A. T. Matheson, P. B. Moore, and H. Noller, eds.), pp. 527–539. ASM Press, Washington, DC.

Kiel, M. C., Raj, V. S., Kaji, H., and Kaji, A. (2003). Release of ribosome-bound ribosome recycling factor by elongation factor G. *J. Biol. Chem.* **278**, 48041–48050.

Kim, K. S., Kang, J. O., Eun, C. S., Han, D. S., and Choi, T. Y. (2002). Mutations in the 23S rRNA gene of *Helicobacter pylori* associated with clarithromycin resistance. *J. Korean Med. Sci.* **17**, 599–603.

Kirsebom, L. A., Amons, R., and Isaksson, L. A. (1986). Primary structure of mutationally altered ribosomal protein L7/L12 and their effects on cellular growth and translational accuracy. *Eur. J. Biochem.* **156**, 669–675.

Kisselev, L., Ehrenberg, M., and Frolova, L. (2002). Termination of translation: Interplay of mRNA, rRNAs and release factors? *EMBO J.* **22**, 175–182.

Kloss, P., Xiong, L., Shinabarger, D. L., and Mankin, A. S. (1999). Resistance mutations in 23S rRNA identify the site of action of the protein synthesis inhibitor linezolid in the ribosomal peptidyl transferase center. *J. Mol. Biol.* **294**, 93–101.

Kolesnikov, A. V., and Gudkov, A. T. (2003). Mutational analysis of the functional role of the loop region in the elongation factor G fourth domain in the ribosomal translocation. *Mol. Biol. (Mosk)* **37**, 719–725.

Koosha, H., Cameron, D., Andrews, K., Dahlberg, A. E., and March, P. E. (2000). Alterations in the peptidyltransferase and decoding domains of ribosomal RNA suppress mutations in the elongation factor G gene. *RNA* **6**, 1166–1173.

Lancaster, L., and Noller, H. F. (2005). Involvement of 16S rRNA nucleotides G1338 and A1339 in discrimination of initiator tRNA. *Mol. Cell* **20**, 623–632.

Larigauderie, G., Laalami, S., Nyengaard, N. R., Grunberg-Manago, M., Cenatiempo, Y., Mortensen, K. K., and Sperling-Petersen, H. U. (2000). Mutation of Thr445 and Ile500 of initiation factor 2 G-domain affects *Escherichia coli* growth rate at low temperature. *Biochimie* **82**, 1091–1098.

LaTeana, A., Brandi, A., O'Connor, M., Freddi, S., and Pon, C. L. (2001). Translation during cold adaptation does not involve mRNA-rRNA base pairing through the downstream box. *RNA* **6**, 1393–1402.

Laursen, B. S., Sørensen, H. P., Mortensen, K. K., and Sperling-Petersen, H. U. (2005). Initiation of protein synthesis in bacteria. *Microbiol. Mol. Biol. Rev.* **69**, 101–123.

Lee, K., Varma, S., SantaLucia, J., and Cunningham, P. R. (1997). *In vivo* determination of RNA structure-function relationships: Analysis of the 790 loop in ribosomal RNA. *J. Mol. Biol.* **269**, 732–743.

Lee, K., Holland-Staley, C. A., and Cunningham, P. R. (2001). Genetic approaches to studying protein synthesis: Effects of mutations at Psi516 and A535 in *Escherichia coli* 16S rRNA. *J. Nutr.* **131**, 2994S–3004S.

Leonov, A. A., Sergiev, P. V., Bogdanov, A. A., Brimacombe, R., and Dontsova, O. A. (2003). Affinity purification of ribosomes with a lethal G2655C mutation in 23S rRNA that affects the translocation. *J. Biol. Chem.* **278**, 25664–25670.

Leontis, N. B., Altman, R. B., Berman, H. M., Brenner, S. E., Brown, J. W., Engelke, D. R., Harvey, S. C., Holbrook, S. R., Jossinet, F., Lewis, S. E., Major, F., Mathews, D. H., *et al.* (2006). The RNA ontology consortium: An open invitation to the RNA community. *RNA* **12**, 533–541.

Liiv, A., and Remme, J. (1998). Base-pairing of 23S rRNA ends is essential for ribosomal large subunit assembly. *J. Mol. Biol.* **276**, 537–545.

Liiv, A., Karitkina, D., Maivali, U., and Remme, J. (2005). Analysis of the function of *E. coli* 23S rRNA helix-loop 69 by mutagenesis. *BMC Mol. Biol.* **6**, 18.

Liljas, A. (2004). "Structural Aspects of Protein Synthesis." World Scientific, Singapore.

Liljas, A., Kristensen, O., Laurberg, M., Al-Karadaghi, S., Gudkov, A., Martemyanov, K., Hughes, D., and Nagaev, I. (2000). The states, conformational dynamics, and fusidic acid-resistant mutants of elongation factor G. *In* "The Ribosome: Structure, Function, Antibiotics and Cellular

Interactions" (R. A. Garrett, S. R. Douthwaite, A. Liljas, A. T. Matheson, P. B. Moore, and H. Noller, eds.), pp. 359–365. ASM Press, Washington, DC.

Liu, M., and Douthwaite, S. (2002). Resistance to the macrolide antibiotic tylosin is conferred by single methylations at 23S rRNA nucleotides G748 and A2058 acting in synergy. *Proc. Natl. Acad. Sci. USA* **99**, 14658–14663.

Lucier, T. S., Heitzman, K., Liu, S.-K., and Hau, P.-C. (1995). Transition mutations in the 23S rRNA of erythromycin-resistant isolates of *Mycoplasma pneumoniae*. *Antimicrob. Agents Chemother.* **39**, 2770–2773.

Macvanin, M., Bjorkman, J., Eriksson, S., Rhen, M., Trimasson, D. I., and Hughes, D. (2003). Fusidic acid-resistant mutants of *Salmonella enterica serovar typhimurium* with low fitness *in vivo* are defective in RpoS induction. *Antimicrob. Agents Chemother.* **47**, 3743–3749.

Maguire, B. A., Beniaminov, A. D., Ramu, H., Mankin, A. S., and Zimmermann, R. A. (2005). A protein component at the heart of an RNA machine: The importance of protein L27 for the function of the bacterial ribosome. *Mol. Cell* **20**, 427–435.

Malbruny, B., Canu, A., Bozdogan, B., Fantin, B., Zarrouk, V., Dutka-Malen, S., Feger, C., and Leclercq, R. (2002a). Resistance to quinupristin-dalfopristin due to mutation of L22 ribosomal protein in *Staphylococcus aureus*. *Antimicrob. Agents Chemother.* **46**, 2200–2207.

Malbruny, B., Nagai, K., Coquemont, M., Bozdogan, B., Andresevic, A. T., Hupkova, H., Leclercq, R., and Appelbaum, P. C. (2002b). Resistance to macrolides in clinical isolates of *Streptococcus pyogenes* due to ribosomal mutations. *J. Antimicrob. Chemother.* **49**, 935–939.

Mankin, A. S. (1997). Pactamycin resistance mutations in functional sites of 16S rRNA. *J. Mol. Biol.* **274**, 8–15.

Matassova, N. B., Rodnina, M. V., and Wintermeyer, W. (2001). Elongation factor G-induced structural change in helix 34 of 16S rRNA related to translocation on the ribosome. *RNA* **7**, 1879–1885.

McGarry, K. G., Walker, S. E., Wang, H., and Fredrick, K. (2005). Destabilization of the P Site codon-anticodon helix results from movement of tRNA into the P/E hybrid state within the ribosome. *Mol. Cell* **20**, 613–622.

Merali, Z., and Giles, J. (2005). Databases in peril. *Nature* **435**, 1010–1011.

Meskauskas, A., Petrov, A. N., and Dinman, J. D. (2005). Identification of functionally important amino acids of ribosomal protein L3 by saturation mutagenesis. *Mol. Cell. Biol.* **25**, 10863–10874.

Mikkola, R., and Kurland, C. G. (1991). Evidence for demand-regulation of ribosome accumulation in. *E. coli. Biochimie* **12**, 1551–1556.

Mohr, D., Wintermeyer, W., and Rodnina, M. V. (2000). Arginines 29 and 59 of elongation factor G are important for GTP hydrolysis or translocation on the ribosome. *EMBO J.* **19**, 3458–3464.

Moore, P. B. (2005). A ribosomal coup: *E. coli* at last. *Science* **310**, 793–795.

Moore, P. B., and Steitz, T. A. (2005). The ribosome revealed. *Trends Biochem. Sci.* **30**, 281–283.

Moore, P. B., and Steitz, T. A. (2006). The roles of RNA in the synthesis of protein. "The RNA World: The Nature of Modern RNA Suggests a Prebiotic RNA World," 3rd edn., pp. 257–285. Cold Spring Harbor Laboratory Press. Cold Spring Harbor, NY.

Morosyuk, S. V., SantaLucia, J., and Cunningham, P. R. (2001). Structure and function of the conserved 690 hairpin in *Escherichia coli* 16S ribosomal RNA. III. Functional analysis of the 690 loop. *J. Mol. Biol.* **307**, 213–228.

Nakamura, Y., Kawazu, Y., Uno, M., Yoshimura, K., and Ito, K. (2000). Genetic probes to bacterial release factors: tRNA mimicry hypothesis and beyond. *In* "The Ribosome: Structure, Function, Antibiotics and Cellular Interactions" (R. A. Garrett, S. R. Douthwaite, A. Liljas, A. T. Matheson, P. B. Moore, and H. Noller, eds.), pp. 519–526. ASM Press, Washington, DC.

Noller, H. F. (2005). RNA Structure: Reading the ribosome. *Science* **309**, 1508–1514.

Noller, H. F. (2006). Evolution of ribosomes and translation from an RNA world. "The RNA World: The Nature of Modern RNA Suggests a Prebiotic RNA World," 3rd edn., pp. 287–307. Cold Spring Harbor Laboratory Press. Cold Spring Harbor, NY.

Noller, H. F., Hoang, L., and Fredrick, K. (2005). The 30S ribosomal P site: A function of 16S rRNA. *FEBS Lett.* **579**, 855–858.

O'Connor, M., Gregory, S. T., and Dahlberg, A. E. (2004). Multiple defects in translation associated with altered ribosomal protein L4. *Nucleic Acids Res.* **32**, 5750–5756.

Okada, T., Wower, I. K., Wower, J., Zwieb, C. W., and Kimura, M. (2004). Contribution of the second OB fold of ribosomal protein S1 from *Escherichia coli* to the recognition of TmRNA. *Biosci. Biotechnol. Biochem.* **68**, 2319–2325.

Pereyre, S., Guyot, C., Renaudin, H., Charron, A., Bebear, C., and Bebear, C. M. (2004). *In vitro* selection and characterization of resistance to macrolides and related antibiotics in *Mycoplasma pneumoniae*. *Antimicrob. Agents Chemother.* **48**, 460–465.

Pfister, P., Hobbie, S., Vicens, Q., Bottger, E. C., and Westhof, E. (2003a). The molecular basis for A-site mutations conferring aminoglycoside resistance: Relationship between ribosomal suscep-tibility and X-ray crystal structures. *Chembiochem* **4**, 1078–1088.

Pfister, P., Risch, M., Brodersen, D. E., and Bottger, E. C. (2003b). Role of 16S rRNA helix 44 in ribosomal resistance to hygromycin. *Antimicrob. Agents Chemother.* **47**, 1496–1502.

Pfister, P., Hobbie, S., Bruell, C., Corti, N., Vasella, A., Westhof, E., and Bottger, E. C. (2004a). Mutagenesis of 16S rRNA C1409-G1491 base-pair differentiates between 6'OH and 6'NH$_3^+$ aminoglycosides. *J. Mol. Biol.* **346**, 467–475.

Pfister, P., Jenni, S., Poehlsgaard, J., Thomas, A., Douthwaite, S., Ban, N., and Bottger, E. C. (2004b). The structural basis of macrolide-ribosome binding assessed using mutagenesis of 23S rRNA positions 2058 and 2059. *J. Mol. Biol.* **342**, 1569–1581.

Pfister, P., Corti, N., Hobbie, S., Bruell, C., Zarivach, R., Yonath, A., and Böttger, E. C. (2005). 23S rRNA base pair 2057–2611 determines ketolide susceptibility and fitness cost of the macrolide resistance mutation 2058A > G. *Proc. Natl. Acad. Sci. USA* **102**, 5180–5185.

Poehlsgaard, J., Pfister, P., Bottger, E. C., and Douthwaite, S. (2005). Molecular mechanisms by which rRNA mutations confer resistance to clindamycin. *Antimicrob. Agents Chemother.* **49**, 1553–1555.

Polacek, N., Gaynor, M., Yassin, A., and Mankin, A. S. (2001). Ribosomal peptidyl transferase can withstand mutations at the putative catalytic nucleotide. *Nature* **411**, 498–501.

Prammananan, T., Sander, P., Brown, B. A., Frischkkorn, K., Onyi, G. O., Zhang, Y., Bottger, E. C., and Wallace, R. J. (1998). A single 16S ribosomal RNA substitution is responsible for resistance to amikacin and other 2-deoxystreptamine aminoglycosides in *Mycobacterium abscessus* and *Mycobacterium chelonae*. *J. Infect. Dis.* **177**, 1573–1581.

Prunier, A.-L., Malbruny, B., Tande, D., Picard, B., and Leclercq, R. (2002). Clinical isolates of *Staphylococcus aureus* with ribosomal mutations conferring resistance to macrolides. *Antimicrob. Agents Chemother.* **46**, 3054–3056.

Rackham, O., Wang, K., and Chin, J. W. (2006). Functional epitopes at the ribosome subunit interface. *Nat. Chem. Biol.* **2**, 254–258.

Recht, M. I., and Puglisi, J. D. (2001). Aminoglycoside resistance with homogeneous and hetero-geneous populations of antibiotic-resistant ribosomes. *Antimicrob. Agents Chemother.* **45**, 2414–2419.

Robert, F., and Brakier-Gingras, L. (2003). A functional interaction between ribosomal proteins S7 and S11 within the bacterial ribosome. *J. Biol. Chem.* **278**, 44913–44920.

Rodnina, M. V., Pape, T., Sabelsbergh, A., Mohr, D., Matassova, N. B., and Wintermeyer, W. (2000). Mechanisms of partial reactions of the elongation cycle catalyzed by elongation factors Tu and G. *In* "The Ribosome: Structure, Function, Antibiotics and Cellular Interactions"

(R. A. Garrett, S. R. Douthwaite, A. Liljas, A. T. Matheson, P. B. Moore, and H. Noller, eds.), pp. 301–317. ASM Press, Washington, DC.

Rodnina, M. V., Gromadski, K. B., Kothe, U., and Wieden, H.-J. (2005). Recognition and selection of tRNA in translation. *FEBS Lett.* **579,** 938–942.

Rodriguez-Correa, D., and Dahlberg, A. E. (2004). Genetic evidence against the 16S ribosomal RNA helix 27 conformational switch model. *RNA* **10,** 28–33.

Roll-Mecak, A., Shin, B.-S., Dever, T. E., and Burley, S. K. (2001). Engaging the ribosome: Universal IFs of translation. *Trends Biochem. Sci.* **26,** 705–709.

Ross, J. I., Eady, E. A., Cove, J. H., and Cunliffe, W. J. (1998). 16S rRNA mutation associated with tetracycline resistance in a gram positive bacterium. *Antimicrob. Agents Chemother.* **42,** 1702–1705.

Sander, P., Belova, L., Kidan, Y. G., Pfister, P., Mankin, A. S., and Bottger, E. C. (2002). Ribosomal and non-ribosomal resistance to oxalolidinones: Species-specific idiosyncrasy of ribosomal alterations. *Mol. Microbiol.* **46,** 1295–1304.

Schlunzen, F., Harms, J. M., Francheschi, F., Hansen, H. A. S., Bartels, H., Zarivach, R., and Yonath, A. (2003). Structural basis for the antibiotic activity of ketolides and azalides. *Structure* **11,** 329–338.

Schuwirth, B. S., Borovinskaya, M. A., Hau, C. W., Zhang, W., Vila-Sanjurjo, A., Holton, J. M., and Cate, J. H. (2005). Structures of the bacterial ribosome at 3.5 Å resolution. *Science* **310,** 827–834.

Serganov, A., Polonskala, A., Ehresmann, B., Ehresmann, C., and Patel, D. J. (2003). Ribosomal protein S15 represses its own translation via adaptation of an rRNA-like fold within its mRNA. *EMBO J.* **22,** 1898–1908.

Sergiev, P. V., Bogdanov, A. A., Dahlberg, A. E., and Dontsova, O. (2000). Mutations at position A960 of E. coli 23S ribosomal RNA influence the structure of 5S ribosomal RNA and the peptidyltransferase region of 23S ribosomal RNA. *J. Mol. Biol.* **299,** 379–389.

Sergiev, P. V., Lesnyak, D. V., Kiparisov, S. V., Burakovsky, D. E., Leonov, A. A., Bogdanov, A. A., Brimacombe, R., and Dontsova, O. A. (2005). Function of the ribosomal E-site: A mutagenesis study. *Nucleic Acids Res.* **33,** 6048–6056.

Sharer, J. D., Koosha, H., Church, W. B., and March, P. E. (1999). The function of conserved amino acid residues adjacent to the effector domain in elongation factor G. *Proteins* **37,** 293–302.

Spahn, C. M., Gomez-Lorenzo, M. G., Grassucci, R. A., Jorgensen, R., Andersen, G. R., Beckmann, R., Penczek, P. A., Ballesta, J. P., and Frank, J. (2004). Domain movements of elongation factor eEF2 and the eukaryotic 80S ribosome facilitate tRNA translocation. *EMBO J.* **10,** 1008–1019.

Springer, B., Kidan, Y. G., Prammananan, T., Ellrott, K., Bottger, E. C., and Sander, P. (2001). Mechanisms of streptomycin resistance: Selection of mutations in the 16S rRNA gene conferring resistance. *Antimicrob. Agents Chemother.* **45,** 2877–2884.

Suzuki, Y., Katsukawa, C., Tamaru, A., Abe, C., Makino, M., Mizuguchi, Y., and Taniguchi, H. (1998). Detection of kanamycin-resistant *Mycobacterium tuberculosis* by identifying mutations in the 16S rRNA gene. *J. Clin. Microbiol.* **36,** 1220–1225.

Tait-Kamradt, A., Davies, T., Appelbaum, P. C., Depardieu, F., Courvalin, P., Petitpas, J., Wondrack, L., Walker, A., Jacobs, M. R., and Sutcliffe, J. (2000a). Two new mechanisms of macrolide resistance in clinical strains of *Streptococcus pneumoniae* from Eastern Europe and North America. *Antimicrob. Agents Chemother.* **44,** 3395–3401.

Tait-Kamradt, A., Davies, T., Cronan, M., Jacobs, M. R., Appelbaum, P. C., and Sutcliffe, J. (2000b). Mutations in 23S rRNA and ribosomal protein L4 account for resistance in pneumococcal strains selected *in vitro* by macrolide passage. *Antimicrob. Agents Chemother.* **44,** 2118–2125.

Takyer, S., Hickerson, R. P., and Noller, H. F. (2005). mRNA helicase activity of the ribosome. *Cell* **120,** 49–58.

Tanaka, I., Nakagawa, A., Nakashima, T., Taniguchi, M., Hosaka, H., and Kimura, M. (2000). Structure and evolution of the 23S rRNA binding domain of protein L2. *In* "The Ribosome: Structure, Function, Antibiotics and Cellular Interactions" (R. A. Garrett, S. R. Douthwaite, A. Liljas, A. T. Matheson, P. B. Moore, and H. Noller, eds.), pp. 85–92. ASM Press, Washington, DC.

Thompson, J., Kim, D. F., O'Connor, M., Lieberman, K. R., Bayfield, M. A., Gregory, S. T., Green, R., Noller, H. F., and Dahlberg, A. E. (2001). Analysis of mutations at residues A2451 and G2447 of 23S rRNA in the peptidyltransferase active site of the 50S ribosomal subunit. *Proc. Natl. Acad. Sci. USA* **98**, 9002–9007.

Trieber, C. A., and Taylor, D. E. (2002). Mutations in the 16S rRNA genes of *Helicobacter pylori* mediate resistance to tetracycline. *J. Bacteriol.* **184**, 2131–2140.

Triman, K. (1995). Mutational analysis of 23S rRNA structure and function in *Escherichia coli*. *Adv. Genet.* **33**, 1–39.

Triman, K. (1999). Mutational analysis of 16S rRNA structure and function in *Escherichia coli*. *Adv. Genet.* **41**, 157–195.

Tsagkalia, A., Leontiadou, F., Xaplanteri, M. A., Papadopoulos, G., Kalpaxis, D. L., and Choli-Papadopoulou, T. (2005). Ribosomes containing mutants of L4 ribosomal protein from *Thermus thermophilus* display multiple defects in ribosomal functions and sensitivity against erythromycin. *RNA* **11**, 1633–1639.

Vila-Sanjurjo, A., and Dahlberg, A. E. (2001). Mutational analysis of the conserved bases C1402 and A1500 in the center of the decoding domain of *Escherichia coli* 16S rRNA reveals an important tertiary interaction. *J. Mol. Biol.* **308**, 457–463.

Vila-Sanjurjo, A., Squires, C. L., and Dahlberg, A. E. (1999). Isolation of kasugamycin resistant mutants in the 16S ribosomal RNA of *Escherichia coli*. *J. Mol. Biol.* **293**, 1–8.

Walsh, F., Willcock, J., and Amyes, S. (2003). High-level telithromycin resistance in laboratory-generated mutants of *Streptococcus pneumoniae*. *J. Antimicrob. Chemother.* **52**, 345–353.

White, S. A., Hoeger, M., Schweppe, J. J., Shillingsford, A., Shipilov, V., and Zarutskie, J. (2004). Internal loop mutations in the ribosomal protein L30 binding site of the yeast L30 RNA transcript. *RNA* **10**, 369–377.

Wilson, D. N., Dalphin, M. E., Pel, H. J., Major, L. L., Mansell, J. B., and Tate, W. P. (2000). Factor-mediated termination of protein synthesis: A welcome return to the mainstream of translation. *In* "The Ribosome: Structure, Function, Antibiotics and Cellular Interactions" (R. A. Garrett, S. R. Douthwaite, A. Liljas, A. T. Matheson, P. B. Moore, and H. Noller, eds.), pp. 495–508. ASM Press, Washington, DC.

Xiong, L., Shah, S., Mauvais, P., and Mankin, A. S. (1999). A ketolide resistance mutation in domain II of 23S rRNA reveals the proximity of hairpin 35 to the peptidyl transferase centre. *Mol. Microbio.* **31**, 633–639.

Xiong, L., Kloss, P., Douthwaite, S., Andersen, N. M., Swaney, S., Shinabarger, D. L., and Mankin, A. S. (2000). Oxazolidinone resistance mutations in 23S rRNA of *Escherichia coli* reveal the central region of domain V as the primary site of drug action. *J. Bacteriol.* **182**, 5325–5331.

Xiong, L., Polacek, N., Sander, P, Bottger, E. C., and Mankin, A. S. (2001). pKa of adenine 2451 in the ribosomal peptidyl transferase center remains elusive. *RNA* **7**, 1365–1369.

Xiong, L., Korkhin, Y., and Mankin, A. S. (2005). Binding site of the bridged macrolides in the *Escherichia coli* ribosome. *Antimicrob. Agents Chemother.* **49**, 281–288.

Xu, W., Pagel, F. T., and Murgola, E. J. (2002). Mutations in the GTPase center of *Escherichia coli* 23S rRNA indicate release factor 2-interactive sites. *J. Bacteriol.* **184**, 1200–1203.

Yassin, A., Fredrick, K., and Mankin, A. S. (2005). Deleterious mutations in small subunit ribosomal RNA identify functional sites and potential targets for antibiotics. *Proc. Natl. Acad. Sci. USA* **102**, 16620–16625.

Youngman, E. M., Brunelle, J. L., Kochaniak, A. B., and Green, R. (2004). The active site of the ribosome is composed of two layers of conserved nucleotides with distinct roles in peptide bond formation and peptide release. *Cell* **28,** 589–599.

Zeidler, W., Egle, C., Ribeiro, S., Wagner, A., Katunin, V., Kreutzer, R., Rodnina, M., Wintermeyer, W., and Sprinzl, M. (1995). Site-directed mutagenesis of *Thermus thermophilus* elongation factor Tu. *Eur. J. Biochem.* **229,** 596–604.

Zhang, S., Ryden-Aulin, M., Kirsebom, L. A., and Isaksson, L. A. (1994). Genetic implication for an interaction between release factor one and ribosomal protein L7/L12 *in vivo. J. Mol. Biol.* **242,** 614–618.

Zimmermann, R. A., Alimov, I., Uma, K., Wu, H., Wower, I., Nikonowicz, E. P., Drygin, D., Dong, P., and Jiang, L. (2000). How ribosomal proteins and RNA recognize one another. *In* "The Ribosome: Structure, Function, Antibiotics and Cellular Interactions" (R. A. Garrett, S. R. Douthwaite, A. Liljas, A. T. Matheson, P. B. Moore, and H. Noller, eds.), pp. 93–104. ASM Press, Washington, DC.

5

Application of Genomics to Molecular Breeding of Wheat and Barley

Rajeev K. Varshney,* Peter Langridge,† and Andreas Graner‡

*International Crops Research Institute for the Semi-Arid Tropics (ICRISAT)
Patancheru 502 324, A.P., India
†Australian Centre for Plant Functional Genomics (ACPFG)
University of Adelaide, Waite Campus, PMB 1 Glen Osmond
SA 5064, Australia
‡Institute of Plant Genetics and Crop Plant Research (IPK)
D06466 Gatrersleben, Germany

Advances in Genetics, Vol. 58
0065-2660/07 $35.00
DOI: 10.1016/S0065-2660(06)58005-8

ABSTRACT

In wheat and barley, several generations of selectable molecular markers have been included in the genetic maps; and a large number of qualitative and quantitative traits were located in the genomes, some of which are being routinely selected in marker-assisted breeding programs. In recent years, a large number of expressed sequence tags (ESTs) have been generated for wheat and barley that have been used for development of functional molecular markers, preparation of transcript maps, and construction of cDNA arrays. These functional genomic resources combined together with new approaches such as expression genetics, association mapping, allele mining, and informatics (bioinformatic tools) possess potential to identify genes responsible for a trait and their deployment in practical plant breeding. High costs currently limit the implementation of functional genomics in breeding programs. The potential applications together with some examples as well as challenges for applying genomics research in breeding activities are discussed. Genomics research will continue to enhance the efficiency and precision for crop improvement but will not replace conventional breeding and evaluation methods.   © 2007, Elsevier Inc.

I. INTRODUCTION

Wheat (*Triticum* spp.) and barley (*Hordeum* spp.) belong to the *Poaceae*, the largest family within the monocotyledonous plants; it includes other major cereal crops of the world such as maize, rice, and rye, as well as important forage grasses such as Ryegrass, Fescue, and Kentucky bluegrass. Among the food crops, wheat and barley are important sources of energy and proteins for the world population and are cultivated over a wide range of climatic regions. Both wheat and barley are among the most extensively studied crop species, particularly in the area of cytogenetics. An extensive catalogue of genetic and cytogenetic stocks such as aneuploid lines, deletion stocks, translocation lines, and so on is available for these crop species (Varshney *et al.*, 2004a,b, 2006a). While barley (*H. vulgare*) is a self-pollinating diploid with $2n = 2x = 14$ chromosomes (H genome), wheat has diploid ($2n = 2x = 14$), tetraploid ($2n = 4x = 28$), and hexaploid ($2n = 6x = 42$) species. However, most modern wheat varieties are hexaploid (*T. aestivum*), described as "common" or "bread" wheat and valued for bread making. Bread wheat is an allopolyploid containing the three distinct but genetically related (homoeologous) genomes—A, B, and D. Although both wheat (hexaploid) and barley are characterized by large genome size with 18,000 and 5000 Mb, respectively, more than 80% of the genome consists of repetitive DNA sequences (Schulman *et al.*, 2004). Such large genomes with the repeated sequences make both genome analysis and crop improvement a challenging task (Langridge *et al.*, 2001).

In recent years, however, due to advances in the area of genetics and genomics, significant progress has been made, and high density molecular genetic as well as physical maps (cytogenetic stocks-based) have become available for wheat and barley. Molecular markers are increasingly being used to tag genes or quantitative trait loci (QTLs) of agronomic importance, offering the possibility of their use in marker-assisted selection (MAS) for breeding (Jahoor *et al.*, 2004; Varshney *et al.*, 2004b, 2006a). In addition to their use in MAS, molecular marker maps have proven to be instrumental resources for the isolation of genes via map-based cloning (Stein and Graner, 2004) and comparative mapping studies in cereal species (see Devos, 2005; Devos and Gale, 2000). Moreover, comprehensive resources, including largest sets of expressed sequence tags (ESTs), bacterial artificial chromosome (BAC) libraries, and DNA arrays, have been developed to facilitate a systematic exploration of the corresponding genomes on the structural and functional levels (Close *et al.*, 2004; Zhang *et al.*, 2004a,b). In this chapter, we review recent progress related to the applications and potential of genomics research in molecular breeding of wheat and barley. Significant emphasis has been laid on the impact of functional genomics and other recent approaches such as association mapping and genetical genomics applied to wheat and barley breeding.

II. MOLECULAR MARKERS AND MARKER-ASSISTED BREEDING

The identification and utilization of genetic variation form the basis of plant breeding. During the process of breeding new varieties, the breeder needs to make decisions at several key points, such as the identification of the most appropriate parents for crosses and the selection of the most desirable individuals among the progeny of the cross (Langridge and Chalmers, 2004). To assess the efficiency of the breeding and selection process, a key issue in any plant-breeding program is the number of lines carried through the evaluation and selection phases. For large wheat- and barley-breeding programs, hundreds of thousands of lines are often required to produce a new variety. In order to save the costs related to running extensive field trials and carrying out the evaluation of some traits, for example, components of grain quality and yield stability, molecular markers provide the opportunities for replacing the expensive and often unreliable bioassays in a cost-effective manner (Koebner *et al.*, 2001). Molecular markers are now widely used to track loci and genome regions in many wheat- and barley-breeding programs, as molecular markers tightly linked with a large number of agronomic and disease resistance traits are available in these species (Gupta *et al.*, 1999; Jahoor *et al.*, 2004; Tuberosa and Salvi, 2004;

Varshney *et al.*, 2006a). In fact, a variety of molecular markers, including restriction fragment length polymorphism (RFLP), random amplification of polymorphic DNA (RAPD), amplified fragment length polymorphism (AFLP), and microsatellite or simple sequence repeat (SSR), have been used for gene tagging and QTL analysis. However, the consensus is that SSRs are presently best suited for the use in marker-assisted breeding (Gupta and Varshney, 2000; Gupta *et al.*, 2002b). RFLP is not readily adapted to high sample throughput, and RAPD assays are not sufficiently reproducible or transferable between laboratories. While both SSRs and AFLPs are efficient in identifying polymorphisms, SSRs are more readily automated (Shariflou *et al.*, 2003). Although AFLPs can be principally converted into simple PCR assays (STS), this conversion can become complicated in large genome templates, as individual bands are often composed of multiple fragments (Carter *et al.*, 2003; Shan *et al.*, 1999). Although use of rare cutter restriction enzyme, such as *Pst*I, in AFLP can increase the frequency of single copy AFLPs; but on the other hand this technique has the risk of detecting only methylation polymorphisms, which may not be stable within or between genotypes (Pellio *et al.*, 2005). Other classes of molecular markers, that are, single nucleotide polymorphisms (SNPs) or single feature polymorphisms (SFPs) are being developed and integrated to genetic maps (Cui *et al.*, 2005; Kota *et al.*, 2003; Rostoks *et al.*, 2005). The inclusion of many molecular markers, especially microsatellite markers on genetic maps of wheat and barley (Table 5.1), will ease their use for marker-assisted breeding.

A. Functional molecular markers

Most of the molecular markers developed, as mentioned above, have been designed from genomic DNA sequences, and therefore they could belong to either transcribed or nontranscribed regions of the genome. Such markers sample genetic variation in the genome more or less randomly and are sometimes referred to as "neutral" or "random" markers (RMs). However, over the last few years, functionally characterized genes, ESTs, and genome-sequencing projects have facilitated the development of molecular markers from the transcribed regions of the genome. Among the more important and popular molecular markers that can be developed from ESTs are SNPs (Rafalski, 2002), SSRs (Varshney *et al.*, 2002, 2005a), or COS (conserved othologous set—the markers that can be used across species, as sequences for such markers are highly conserved; Fulton *et al.*, 2002; Rudd *et al.*, 2005). Putative functions can be deduced for the markers derived from ESTs/genes using homology searches (BLASTX) with protein databases (e.g., NR-PEP, SWISSPROT, and so on). Therefore molecular markers, generated by utilizing (gene) sequence data, are known as "functional markers" (FMs; Andersen and Lubberstedt, 2003). FMs have some advantages compared with RMs as they are completely linked to

Table 5.1. Available SSR Markers in Wheat and Barley

Populations	Number and type[a] of SSR loci mapped	Designation of SSR	References
Wheat			
ITMI RILs (W7984 × Opata85)	279 gSSR	gwm	Röder et al. (1998)
F2s (Chinese Spring × Synthetic)	53 gSSR	psp	Stephenson et al. (1998)
ITMI RILs (W7984 × Opata85)	65 gSSR	gdm	Pestsova et al. (2000)
ITMI RILs (W7984 × Opata85, deletion lines)	168 gSSR	barc	Song et al. (2002)
ITMI RILs (W7984 × Opata85)	66 gSSR	wmc	Gupta et al. (2002a)
ITMI RILs (W7984 × Opata85)	84 gSSR	cfd	Guyomarc'h et al. (2002)
ITMI RILs (W7984 × Opata85)	22 eSSR	DupW	Eujayl et al. (2002)
Consensus map based on four mapping populations (W7984 × Opata85, RL4452 × AC Domain, Wuhan × Maringa, Superb × BW278)	1108 gSSR	wmc, gwm, gdm, barc, cfa, cfd	Somers et al. (2004)
RILs (W7984 × Opata85; Wenmai6 × Shanhongmai), DHs (Lumai14 × Hanxuan10)	101 eSSR	Cwm or GeneName-SSR	Gao et al. (2004)
ITMI RILs (W7984 × Opata85)	149 eSSR	cnl, ksu	Yu et al. (2004a)
RILs (W7984 × Opata85)	126 eSSR	gpw/cfe	Nicot et al. (2004)
RILs (W7984 × Opata85)	48 eSSR	cwem	Peng and Lapitan (2005)
ITMI RILs (W7984 × Opata85)	>600 gSSR	gwm	Röder and Ganal (personal communication)

(Continues)

Table 5.1. (*Continued*)

Populations	Number and type[a] of SSR loci mapped	Designation of SSR	References
Barley			
DHs (Igri × Franka, Steptoe × Morex, Harrington × Morex, Harrington × TR306)	45 gSSR	HVM	Liu et al. (1996)
DHs (*H. vulgare* var Lina × *H. spontaneum* Canada Park)	242 gSSR	Bmac, Bmag, Ebmac, Ebmag, HvGeneName	Ramsay et al. (2000)
F$_2$s (Lerche × BGRC41936), DHs (Igri × Franka)	57 eSSR	HvGeneName	Pillen et al. (2000)
Consensus map based on three mapping populations (Igri × Franka; Steptoe × Morex; OWB Dom × OWB Rec)	185 eSSR	GBM	Thiel et al. (2003), Varshney et al. (2006b)
DHs (Igri × Franka, Steptoe × Morex)	127 gSSR	GBMS	Li et al. (2003)
DHs (*H. vulgare* var Lina × *H. spontaneum* Canada Park), F$_2$ (SuspTrit × Cebada Cepa)	65 eSSR	GBM	Marcel and Niks, Wageningen (personal communication)

[a]gSSR, derived from genomic DNA after isolating from genomic library; eSSR, derived from ESTs after searching ESTs for SSRs.

the corresponding trait allele (Varshney et al., 2005c). Such markers may be derived from the gene responsible for the trait of interest and target the functional polymorphism in the gene, thus allowing selection in different genetic backgrounds without revalidating the marker–QTL allele relationship. Thus, they have also been referred as "perfect markers" or "diagnostic marker" even though different alleles with the same polymorphism (resulting from intragenic recombination, insertion, deletion, or mutation) may produce different phenotypes. A perfect marker allows breeders to track specific alleles within pedigrees and populations and minimize linkage drag—segregation of undesirable segments with gene of interest—flanking that gene.

In recent years, due to emphasis on functional genomics, an excellent collection of ESTs has been developed in wheat (Zhang et al., 2004a) and barley (Zhang et al., 2004b). In terms of numbers, wheat with 600,039 and barley with 419,146 ESTs rank number second and fourth among EST collections for plant species (http://www.ncbi.nlm.nih.gov/dbEST/dbEST_summary.html, dbEST release 102805). Both genetic and deletion stocks (with various sized terminal deletions in individual chromosome arms, useful for sub-arm localization of genes/markers)-based physical maps of wheat and barley have been generated with EST-based markers, for example EST-SSRs (Gao et al., 2004; Nicot et al., 2004; Peng and Lapitan, 2005; Thiel et al., 2003; Yu et al., 2004a; see Table 5.1), EST-SNPs (Kota et al., 2001; Somers et al., 2003), EST-RFLPs (Qi et al., 2004; Varshney et al., 2005c), and EST-CAPS (K. Sato, Japan, personal communication). As a result, high-density transcript maps for barley with over 3000 EST loci (personal communications with K. Sato, Japan; R. Waugh, UK), and wheat with over 16,000 EST loci (Qi et al., 2004), have or will be shortly available. Microarray-based gene-expression data between two genetically different lines can also be utilized to identify single feature polymorphisms (SFPs) for SNP detection in a highly parallel manner (Borevitz et al., 2003) which can be exploited to develop FMs. In fact, in a study using 17 and 19 Affymetrix Gene-Chip expression datasets for two genotypes, >10,000 SFPs have been identified between the two genotypes of barley, a species with a large and complex genome (Rostoks et al., 2005). By using another alternative method, with a smaller number (three) of replicate datasets and a different statistical method (robustified projection pursuit), about 2000 SFPs have been identified in another study of barley (Cui et al., 2005). However, identification of SFPs involves the problem of sensitivity versus selectivity, that is, a large number of putative SNPs could not be confirmed (Rostoks et al., 2005). Furthermore, the development of SNP markers in polyploid crop species such as wheat is complicated by the need to distinguish intragenome from intergenome polymorphisms, referring to the "three genomes" carried by this species: A, B, and D (Powell and Langridge, 2004).

B. Status of marker-assisted breeding

Prior to the deployment of DNA markers in plant breeding, the markers need to be validated, a process in which functionality is tested in a range of genetic backgrounds (Gupta et al., 1999; Langridge and Chalmers, 1998). For instance in wheat, marker validation studies were conduced for QTLs for grain protein content by using near isogenic lines (NILs), the lines that are genetically identical except at one or a few loci (Singh et al., 2001), for Lr10 by using 16 wheat cultivars (Blazkova et al., 2002), for the QTL for Fusarium head blight (FHB) resistance by using the progeny of crosses between the FHB-resistant spring wheat line and five European wheat varieties (Angerer et al., 2003), or NILs from existing breeding populations (Pumphrey and Anderson, 2003) and in germplam collections (Zhou et al., 2003). Similarly markers associated with preharvest sprouting (Kato et al., 2001; Mares and Mrva, 2001), plant height (Ellis et al., 2002), and barley yellow dwarf virus (Ayala et al., 2001) were validated and used for enriching favorable allele frequencies in segregating populations and tracking donor parent alleles during backcrossing (Cakir et al., 2003) in wheat. Molecular markers have also facilitated the pyramiding of multiple disease resistance genes in wheat and barley. For example, Liu et al. (2000) integrated three powdery mildew resistance gene combinations (Pm2 + Pm4a, Pm2 + Pm21, and Pm4a + Pm21) into an elite wheat cultivar called "Yang158." Similarly, marker-assisted backcross introgression of Yd2 gene conferring resistance to barley yellow dwarf virus was conduced in barley (Jefferies et al., 2003).

A particular effort to use MAS in wheat- and barley-breeding programs has been initiated in Australia. More than two dozen loci each in wheat and barley are currently being used in the Australian cereal-breeding programs (Langridge, 2005). For breeding programs in which molecular markers are actively being used, it has been estimated that over half the varieties currently being released have used markers at some stage during the breeding process. For instance, in the South Australian Barley Improvement Program markers were deployed to eliminate defects in elite varieties and a "Sloop type" variety with cereal cyst nematode (CCN) resistance was advanced to commercial release in less than 8 years (Langridge, 2005). Several other new Australian varieties have been developed through the use of markers. While using marker-assisted backcrossing in combination with the production of doubled haploids (DHs), the time from the initial cross to release of the variety has been almost halved when compared with conventional breeding. In fact in conventional breeding programs, 12 years were required on average from the first cross to the release of a wheat variety and 14 years for a malting quality variety in the breeding programs of South Australia. The barley variety "Tango," released in 2000 in the United States, is claimed to be the first barley variety developed by molecular MAS. It contains two QTLs for adult resistance to stripe rust, a character difficult to handle by conventional phenotypic selection (Toojinda et al., 1998). These were transferred into the 1970s variety "Steptoe" via two cycles of RFLP-aided backcrossing (Hayes et al., 2003a). Although "Tango" has a good level of rust resistance, its yield is less than that of its long outclassed recurrent parent, and hence it is seen primarily as a genetically characterized source of resistance rather than as a variety in its own right. Just as for wheat, most of the proposed targets for MAS in barley relate to genes for disease resistance, although for many of these disease-efficient phenotypic screens are available. On the other hand, malting quality represents an important QTL target for MAS breeding in barley, because this complex trait is difficult to score and malting barley attracts a substantial price premium compared to feed barley (Ayoub et al., 2003; Han et al., 1997; Hayes et al., 2003b).

The above examples demonstrate the potential of marker-assisted breeding concepts. Nevertheless, the best possible integration of marker-assisted concepts in a given breeding program depends on a variety of issues, such as the traits under consideration and the availability of closely linked markers, the costs for phenotypic versus MAS as well as the breeding scheme (backcross vs pedigree based approaches). On the basis of a simulation study, the combination of MAS at the BC_1F_1 and haploid stage was identified as the optimal strategy (Kuchel et al., 2005). This study showed that incorporation of marker selection at these two stages not only increased genetic gain over the phenotypic alternative but also actually reduced the overall costs by 40%. Furthermore, as the unit marker assay costs are expected to decrease with the development of automated platforms and high-throughput marker systems, it is anticipated that MAS assays will become increasingly competitive. In a similar way, the deployment of MAS for wheat and barley breeding will benefit from development of additional FMs.

C. Whole-genome breeding

As mentioned above, in several cases molecular makers have been successfully utilized in trait-based breeding. Since extensive genetic information is available about a wide range of traits covering disease resistance, abiotic stress tolerance, and aspects of quality, it is possible, in principle, to target large numbers of loci at once in breeding strategies to manage the entire genome. This process is termed as "whole-genome breeding" (Langridge, 2005). There are several ways in which this can be achieved but it does require major shifts in breeding methodologies and is likely to be specific for each task being addressed. The overall concept is illustrated in Fig. 5.1. On the basis of molecular information derived from a large number of crosses, Jefferies (2000) proposed the concept of developing a genetic ideotype. The ideotype shown in Fig. 5.1 was prepared on the basis of known allele composition at 10 target loci or regions for 5 key varieties namely "Alexis" (European malt quality), "Sahara" (North African

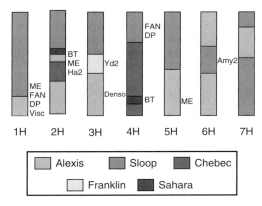

Figure 5.1. Genetic ideotype based around a set of five barley varieties. Each chromosome has been
colored to reflect the desired region. The location of some key loci is indicated (ME, malt
extract; FAN, free amino nitrogen; DP, diastatic power; Visc, wort viscosity; BT, boron
tolerance; Ha2, cereal cyst nematode resistance; Yd2, barley yellow dwarf virus resistance;
Denso, dwarf; Amy2, α-amylase) [adapted from Jefferies (2000)]. (See Color Insert.)

landrace), "Sloop" (Australian malt quality), "Chebec" (well-adapted Australian), and "Franklin" (Australian malt quality). It can be seen that in some cases
there are blocks of loci all with the desirable alleles, the breeders seek to maintain
or transfer these blocks into a new variety, for example the region on linkage
group 1H. This linkage block can be sourced from "Alexis." The linkage group 2H
is complex, and therefore several specific recombination events are required to
bring together the desirable alleles from "Sahara," "Sloop," and "Chebec." For
constructing such a variety, the populations should be designed so as to achieve
the desired structure for each chromosome. Further, the population size can also
be kept manageable by working with the chromosomes (linkage groups) one at a
time and bringing them together only in the final variety.

The whole-genome breeding approach has already been used in
Australian breeding programs. In fact, graphical genotypes developed for DH
lines of a cross between "Alexis" and "Sloop" revealed one individual with the
appropriate configuration for chromosomes 1H, 5H, 6H, and 7H (Fig. 5.1).
Therefore, only remaining chromosomes of the barley genome (2H, 3H, and
4H) needed to be "redesigned," and this could be achieved by selecting lines
with key recombination events and the reassembling all chromosomes in the
final cross (Langridge, 2005). The strategy followed with some modifications was
used to produce the variety "Flagship" that was released in 2004.

The genetic ideotype strategy is only one of many possible whole-
genome breeding approaches that could be applied to wheat and barley improvement. Examples of further applications are underway in the Australian programs

that include work to combine desirable alleles at multiple malt quality loci from European, Japanese, and Canadian lines into a well-adapted and high-yielding Australian background (Vassos et al., 2003) and to correct multiple deficiencies in a high-yielding wheat line (Kuchel et al., 2003). Advances in marker screening and genotyping strategies will help the whole-genome breeding approach, making it possible to be used widely. Furthermore, the shift to association genetic studies (see later) in wheat and barley may provide a view of the key linkage blocks and haplotype structure of these species, which will be crucial in the next generation of whole-genome breeding strategies.

III. GENOMIC RESOURCES AND APPROACHES

Due to significant progress in the area of molecular genetics during the last two decades, enormous genomic resources have been developed for major crop plant species. For example, for wheat and barley, high-density genetic maps, cyto-genetic stocks-, as well as contig-based physical maps, deep coverage large insert such as BAC libraries are available (Gupta and Varshney, 2004). These tools have facilitated isolation of genes or QTLs via map-based cloning approaches leading to sequencing and annotation of large genomic DNA fragments in these species (Salvi and Tuberosa, 2005; Stein and Graner, 2004).

A large collection of sequence data from genome- and EST-sequencing projects, combined with recent advances in the DNA sequence analysis (bio-informatics) and establishment of high throughput assays, have provided the framework for large-scale gene discovery and analysis of DNA sequence variation in plant species. The salient challenge of applied genetics and functional genomics is the correlation between genetic and phenotypic information and the subsequent identification of the genes underlying a trait of interest so that they can be exploited in crop improvement programs.

A. Transcriptome analysis

With the establishment of large-scale EST-programs in several laboratories around the world, a comprehensive resource has been created that provides direct access to genes of wheat and barley. In order to establish an inventory of expressed genes in *Triticeae* species, an international consortium (International Triticeae EST Cooperative (http://wheat.pw.usda.gov/genome/) under the um-brella of International Triticeae Mapping Initiative, http://wheat.pw.usda.gov/ITMI/) was established to trigger the development of a wheat and barley EST database. This effort provided the first serious collection of ESTs and led to other initiatives in the area of wheat and barley genomics.

For wheat, a project entitled "The Structure and Function of Expressed Portion of Wheat Genome" involving 13 laboratories was established in 1999 and funded by the US National Science Foundation (NSF) (http://wheat.pw. usda.gov/NSF/). The objective of this project was to decipher the chromosomal location and biological function of a large set of wheat genes. To this end, a total of 117,510 ESTs (101,912 are 5′ ESTs and 15,605 are 3′ ESTs, as of July 2003) from 20 cDNA libraries were generated (Zhang *et al.*, 2004a). Since ESTs reflect the transcriptional status of the tissue they were derived from, the sequences are inherently redundant (Varshney *et al.*, 2004c). EST clustering is applied to remove the redundancy and to sort the sequences into singletons and sequence clusters (Zhang *et al.*, 2004a). The sum of the numbers of singletons and clusters yields the number of tentative unigenes (TUCs). Computational analysis of the wheat EST dataset, as mentioned above, yielded 18,876 contigs and 23,034 singletons (http://wheat.pw.usda.gov/NSF/curator/assembly.html; Lazo *et al.*, 2004). In addition to these ESTs generated in NSF-sponsored projects, other public laboratories and private organizations, such as the DuPont Corporation, also generated wheat ESTs and submitted them to public databases. As a result 600,039 wheat ESTs are available in the public domain (http://www.ncbi.nlm. nih.gov/dbEST/dbEST_summary.html, dbEST release 102805). A computational analysis of 580,155 wheat ESTs suggested the presence of 44,954 TCs (tentative consensus sequences or consensi) and 77,187 singleton ESTs, as per TIGR Wheat Gene Index Release 10.0, January 14, 2005 (http://www.tigr.org).

Similarly for barley, a large collection of ESTs was generated from over 80 cDNA libraries, covering virtually any tissue and growth stage as well as a series of physiological conditions (e.g., seed development or seed germination at different time intervals). This work has involved several laboratories worldwide including IPK-Gatersleben (Germany), Clemson University (USA), Washington Sate University (USA), SCRI (UK), Okayama University (Japan), and the University of Helsinki (Finland). As a result, 419,146 barley ESTs have become available in public domain as of late October, 2005 (http://www.ncbi. nlm.nih.gov/dbEST/dbEST_summary.html, dbEST release 102805). Cluster analysis of 330,000 ESTs that were available in 2003 resulted in the definition of ca. 33,000 TUCs. A comparison of the available sequence data to 254 well-characterized barley genes from the SWISSPROT database and to 1.2 Mb of annotated BAC-sequence originating from several regions of the barley genome revealed an EST coverage of 87% for the SWISSPROT dataset and 45% for the genomic sequences. Thus, a preliminary estimate of the gene repertoire of barley will lie between 38,000 and 72,000 genes (Graner *et al.*, 2005; Zhang *et al.*, 2004b). However, the complexity of a genome is defined not only by the number of its genes but also by the number of its proteins. The latter may be influenced by alternative splicing, which is a common feature of the human transcriptome (Johnson *et al.*, 2003). In higher plants, alternative splicing may be much more infrequent, since so far only a few cases have been described. Barley EST data revealed that about 4% of the barley genes show alternatively spliced isoforms; a similar figure as was reported for *Arabidopsis* (Brett *et al.*, 2002).

The outcome of cluster analysis (for defining the unigene set) depends on a series of parameters including the average sequence length of an EST, the quality of the sequences, and the contamination of EST data with sequences from other organisms, such as microbes or fungi. Moreover, the result of the analysis is influenced by the stringency of the cluster algorithm. The higher the stringency, the more singletons (which may be due to sequencing errors only) and thus the more unigenes will be defined. Nevertheless, the redundant EST dataset (extensive EST databases prepared from many different tissues) can be used to estimate gene expression levels by measuring the frequency of appearance of specific sequences, employing computational tools such as Digital Differential Display (http://www.ncbi.nlm.nih.gov/UniGene/info_ddd.shtml) or HarvEST (http://harvest.ucr.edu/). An example of the use of wheat ESTs from multiple cDNA libraries to study developmental processes was shown by Ogihara *et al.* (2003). After the analysis of 116,232 ESTs, generated from 10 wheat tissues, they identified correlated expression patterns of genes across the tissues. Furthermore, relationships of gene expression profiles among the 10 wheat tissues were inferred from global gene expression patterns. However, the use of EST databases to study expression profiles is limited by the availability of cDNA libraries used to develop ESTs and by the depth of EST sequencing. There are also problems in tracking genes that may be represented by several partial EST sequences.

B. Functional genomics

Functional genomics involves identification of functional allelic differences conferring an improved phenotype. In such an approach, the objective is to identify a sequence change conferring the improvement. The sequence change can then become the basis for a molecular marker that is specific for that allele. These types of (functional) molecular markers should/will always cosegregate with the trait of interest and should also be polymorphic in any cross, as discussed above. In general, such a marker will often be based on an SNP. Since SNPs can be detected by high-throughput systems, they bear the potential that large numbers of plants can be assayed for a particular allele (Rafalski, 2002). Thus, functional genomics can be linked or associated with plant breeding for crop improvement programs.

For carrying out the functional genomics studies, several techniques or platforms are available that allow the estimation of mRNA abundance for large number of genes simultaneously (Sreenivasulu *et al.*, 2002). The methods include serial analysis of gene expression (SAGE; Velculescu *et al.*, 1995), microarrays (Schena *et al.*, 1995), macroarrays (Desprez *et al.*, 1998), and massively parallel signature sequencing (MPSS; Brenner *et al.*, 2000). SAGE, a logical extension of EST sequencing, can be used to study the expression patterns (Velculescu *et al.*, 1995). An improved variant to the conventional SAGE procedure called "Super-SAGE" was developed by using the type III restriction endonuclease *Eco*P15I for isolating fragments of 26 bp from defined positions of cDNAs (Matsumura *et al.*, 2003). Unfortunately, SAGE or SuperSAGE suffers from several problems. In particular, these experiments require large amounts of RNA and can be very expensive if many samples are to be analyzed, for example from a developmental series. As with MPSS (Brenner *et al.*, 2000; http://www.lynxgen.com/), the signatures generated can be difficult to assign to particular genes when the technique is applied to wheat and barley, where full genome sequences are not available.

Microarrays and macroarrays offer a technique for screening the expression profile of very large numbers of genes simultaneously. Furthermore, macroarrays have the advantage of ease of manufacture and low costs compared to microarrays, but macroarrays do not provide the same level of gene or probe density and specificity (Milligan *et al.*, 2004; Sreenivasulu *et al.*, 2002). On the basis of the available EST information, GeneChip arrays for both barley and wheat genomes have recently been constructed at Affymetrix (http://www.affymetrix.com/). The Barley 1 GeneChip is based on 350,000 high-quality ESTs from 84 cDNA libraries and contains 21,439 nonredundant genes (Close *et al.*, 2004). The Wheat GeneChip array contains 61,127 probe sets representing 55,052 transcripts for all 42 chromosomes in the wheat genome.

DNA arrays have been successfully utilized in many plant species, including cereals such as maize and rice (wheat and barley also), for understanding developmental processes, environmental stress responses, identification, and genotyping of mutations (Aharoni and Vorst, 2002; Milligan *et al.*, 2004; Potokina *et al.*, 2004). However, use of these technologies for applied aspects in plant breeding has been limited as it is not possible to simply estimate a difference in gene expression observed between two phenotypically contrasting lines (for agronomic trait of interest). With the possible exception of NILs (for which lines are genetically identical except a few loci), differential gene expression is not only due to the trait of interest but also due to variation in the genetic background. Therefore, background effects need to be eliminated in order to establish a functional association between the level of gene expression and a given trait. In this context, Potokina *et al.* (2004) established a strategy based on analysis of a representative number of well-described genotypes in terms of various phenotypic parameters for a given trait (malting quality). Subsequently, from a total set of genes which are differentially expressed between the lines by using macro/microarrays, only those genes are extracted whose expression profile accounts for phenotype-based relation between lines. Using this strategy with 10 barley genotypes characterized for 6 malting quality parameters, and a cDNA array with 1400 unigenes, Potokina *et al.* identified between 17 and 30 candidate genes for each of the 6 malting parameters. This set of candidates contained genes that were previously supposed to be related to malting quality (e.g., cysteine proteinase 1), genes hitherto unknown to be related to this trait (such as a gene encoding 70-kDa heat shock protein) and genes of unknown function. Further, the observed linkage of five out of eight mapped candidate genes to known QTLs for malting quality traits underscores the potential usefulness of this approach for the identification of candidate genes for a trait under consideration (Potokina et al., 2006). To reveal spatiotemporal expression patterns, cDNA microarrays containing ~9000 wheat cDNAs were used to monitor gene expression during the first 28 days of grain development following anthesis (Leader et al., 2003). This study revealed 66 differentially regulated genes, which showed sequence similarity to transcription factors. The identified genes can be used for gene-specific marker development and synteny with rice to determine if any of the genes map within regions corresponding to QTL for grain yield or quality traits. Similarly, exploitation of cDNA microarrays is underway to identify genes controlling endosperm development (Shinbata et al., 2003) for studying the Russian wheat aphid (RWA) defense response mechanisms (Botha et al., 2003) and assessment after fungicide application (Pasquer et al., 2003). Such studies (also see Milligan et al., 2004; Sreenivasulu et al., 2004) clearly demonstrate that the functional association strategy can provide an efficient link between functional genomics and plant breeding.

The results from DNA array experiments, however, need to be interpreted with caution. Different microarray platforms (e.g., Affymatrix, Agilent, and Amersham), with the same RNA sample or analysis of the same microarray gene expression data with different bioinformatic tools, may yield different sets of candidate genes (Larkin et al., 2005; Miklos and Maleszka, 2004; Tan et al., 2003). Therefore, to further confirm candidate genes obtained from DNA array analysis reverse-transcription (RT)-PCR may be employed, since it is at least 100-fold more sensitive than DNA arrays in detecting transcripts (Horak and Snyder, 2002). In this regard, Czechowski et al. (2004) developed a real-time RT-PCR-based approach for quantitative measurement of genes including TF genes. Thus, knowledge about where and when TF genes are transcribed and how such transcription is affected by internal and external cues will be valuable in elucidating the specific biological roles of the cognate proteins especially in response to environmental stresses.

C. Expression genetics and eQTLs

Jansen and Nap (2001) outlined the use of gene expression data in QTL analysis and the approach was termed "genetical genomics." However, we prefer to call this approach "expression genetics" as here the expression-profiling data is analyzed in the form of a genetic perspective (Varshney et al., 2005b).

The "expression genetics" or "genetical genomics" approach combines QTL mapping with expression/transcript profiling of individual genes in a segregating (or mapping) population. Genes controlling a particular trait that are differentially expressed between two genotypes are used to record the corresponding expression data on each individual of the mapping population. The level of expression of a gene is treated as a quantitative trait. Assuming that each gene showing transcriptional regulation is mapped within the genome or the species of interest, the expression data is subjected to QTL analysis. The QTLs identified using this approach are popularly called as "e(xpress)QTLs." The dedicated software tool "Expressionview" has also been developed to combine visualization of gene expression data with QTL mapping (Fischer *et al.*, 2003). This tool can be used to utilize gene expression data in the form of QTL analysis to identify the eQTLs.

Because eQTL analysis uses segregating populations, it is possible to determine whether expression of a target gene is regulated in *cis* (mapping of the differentially regulated candidate gene within the eQTL) or *trans* (the candidate gene is located outside the corresponding eQTL). The latter gene product (second order effect) is of interest because more than one QTL can be connected to such a *trans*-acting factor (genes acting on the transcription of other genes) (Schadt *et al.*, 2003). Thus, mapping of eQTLs allows multifactorial dissection of the expression profile of a given mRNA/cDNA, protein, or metabolite into its underlying genetic components, and also allows localization of these components on the genetic map (see Jansen, 2003; Jansen and Nap, 2001). Subsequently eQTL analysis, for each gene or gene product analyzed, can underline the regions of the genome influencing its expression. Furthermore, for plant species such as *Arabidopsis* and rice, for which whole-genome sequences are available, the annotation of those genomic regions, which correspond to an eQTL, will be helpful for the identification of the genes as well as their regulatory sequences (Sreenivasulu *et al.*, 2004).

The genetical genomics approach has already demonstrated its utility in dissection and uncovering the regulatory pathways of complex traits in humans, fruitflies, yeast, and some plants (for reference see Varshney *et al.*, 2005b). This approach holds great potential to pinpoint genes involved in expression of agronomic traits based on the hypothesis that the expression of a quantitative phenotypic trait is a function of the expression level of the underlying genes. The colocalization of candidate genes with QTLs controlling a particular phenotype supports the use of the candidate gene as a potential source for developing "perfect marker(s)" for selecting the phenotype in marker (genomics)-assisted breeding. The availability of large EST collections and GeneChip arrays for genome-wide expression profiling and analytical tools for molecular marker analysis in wheat and barley will accelerate the use of this approach for understanding the genetic control of different agronomic traits for plant breeding.

IV. COMPARATIVE GENOMICS

Significant genomic collinearity in plants has been revealed by comparative genetic mapping, although plant genomes vary tremendously in size, chromosome number, and chromosome morphology. For example, genomic collinearity or conservation of synteny on chromosomes among related species is well known for the *Poaceae* (Ahn *et al.*, 1993; Devos, 2005; Devos and Gale, 1997, 2000), *Solanaceae* (Bonierbale *et al.*, 1988; Tanksley *et al.*, 1992), and between *Arabidopsis* and *Brassica* species (Lagercrantz, 1998). The availability of a large number of ESTs for cereal species including wheat and barley, and the complete genome sequence of rice has allowed sequence comparisons between and among cereal genomes and opened a new area of comparative genomics. The use of DNA sequence-based comparative genomics for evolutionary studies and for transferring information from model species to related large-genome species has revolutionized molecular genetics and breeding strategies for improving these crops (Paterson, 2004). Comparative sequence analysis methods provide cross-reference of genes between the maps of different species, enhance the resolution of comparative maps, help to explain patterns of gene evolution and can be used to identify conserved regions of the genomes, and facilitate interspecies gene cloning (Stein *et al.*, 2000).

Despite the benefits arising from the exploitation of syntenous relationships between genomes, it must be kept in mind that millions of years of genome evolution has left its traces. As a result, marker and gene collinearity are frequently interrupted, and chromosomes represent patchworks of collinear and non-collinear segments. In a comprehensive study, 5780 ESTs physically mapped in wheat chromosome bins (using deletion stocks) were compared by BLAST analysis to 3280 ordered BAC/PAC clones from rice, and numerous chromosomal rearrangements were observed between wheat and rice genomes (La Rota and Sorrells, 2004; Sorrells *et al.*, 2003). In addition, the physical locations of nonconserved regions were not consistent across rice chromosomes. Some wheat ESTs with multiple wheat genome locations were found associated with the nonconserved regions. An average of 35% of the putative single copy genes that were mapped to the most conserved bins matched rice chromosomes other than the one that was most similar overall (Singh *et al.*, 2004). As noted above, interruption of microcollinearity, due to rearrangements, was observed in other studies when extensive comparisons were made across smaller regions between collinear chromosome (/arm) of wheat and rice (Distelfeld *et al.*, 2004; Guyot *et al.*, 2004; Singh *et al.*, 2004).

Sequence comparison (BLASTN analysis) of 974 mapped ESTs from a transcript map of barley with the available 1,369,683 ESTs of six cereal species and 286,255 ESTs of three dicot species showed the presence of barley homologs in all species examined (Varshney *et al.*, 2005c). Among cereals, barley EST-derived markers showed *in silico* transferability of 95.4% to wheat, 70% to rice, 69.7% to maize, and 66.2% to sorghum. A lower transferability of only 38.7%, which was observed for rye, and 3.8% for oats may be attributed to the small datasets (9194 ESTs in rye and 501 ESTs in oats) that were available for analysis and which may be biased regarding the content of conserved sequences. Even significant homology of barley ESTs with an average of 15% ESTs of dicot species suggests that a COS of markers could be developed as demonstrated earlier by Fulton *et al.* (2002).

Despite the frequent observation of disturbed microsynteny between the genomes of *Triticeae* species and rice, at the sequence level, the cross-transfer information from rice (and other cereal genomes) to wheat and barley has greatly facilitated the isolation of genes via map-based cloning approaches in wheat and barley (see Stein and Graner, 2004). In addition, molecular marker resources that are available for wheat and barley have been used to improve the marker density of the genetic maps of other cereal species, such as rye, which lack the corresponding resources (Khlestkina *et al.*, 2004; Varshney *et al.*, 2004c). A summary of selected comparative mapping studies based on molecular marker and/or sequence comparison is given in Table 5.2. These studies will prove useful in comparative mapping among fairly divergent genomes and therefore may also prove useful for taxonomic studies, such as deducing phylogenetic relationships between different genera and species.

V. EXPLOITATION OF NATURAL VARIATION AND ALLELIC DIVERSITY

Existing genetic variation in germplasm collections has been utilized for decades by plant breeders in creating new varieties improved for desired agronomic traits. However, during the process of domestication, the genetic base of crop species has been narrowed (Tanksley and McCouch, 1997). Thus modern breeding is now returning to the wild ancestors of crop plants and employs some of the diversity that was lost during domestication to further improve agricultural performance (Zamir, 2001). Utilization of wild germplasm tends to be complex, as the target loci in the wild material are often transferred in large linkage blocks that adversely affect performance of the adapted parents; the phenomenon is called "linkage drag" (Gur and Zamir, 2004; Tanksley and McCouch, 1997). Although there may be many tools or strategies in modern genetics or genomics that may manage and exploit unused natural-variation potential of wild,

Table 5.2. Comparative Mapping and Genomics Studies Revealing the Syntenic Relationship Between Wheat and Barley and Other Cereal Species

Species	References
Barley, wheat	Dubcovsky et al. (1996), Hernandez et al. (2001), Hohmann et al. (1995), Namuth et al. (1994), Salvo-Garrido et al. (2001), Varshney et al. (2005b), Weng and Lazar (2002)
Barley, rye	Varshney et al. (2005d), Wang et al. (1992)
Barley, rice	Han et al. (1998, 1999), Kilian et al. (1995, 1997), Perovic et al. (2004), Saghai-Maroof et al. (1996), Smilde et al. (2001), Varshney et al. (2005d)
Barley, wheat, rye	Börner et al. (1998), Devos and Gale (1993), Devos et al. (1993b), Gudu et al. (2002)
Barley, wheat, rice	Dunford et al. (1995), Gallego et al. (1998), Kato et al. (2001)
Barley, oat, maize	Yu et al. (1996)
Wheat, rye	Devos et al. (1992, 1993a), Khlestkina et al. (2004)
Wheat, maize	Devos et al. (1994)
Wheat, rice	Francki et al. (2003), Kato et al. (1999), Kurata et al. (1994), Lamoureux et al. (2002), La Rota and Sorrells (2004), Laubin et al. (2003), Li et al. (2004), Liu and Anderson (2003), Sarma et al. (1998, 2000), Singh et al. (2004), Sorrells et al. (2003), Yu et al. (2004b)
Wheat, maize, rice	Ahn et al. (1993), Moore et al. (1995b)
Wheat, maize, oat, rice	Van Deynze et al. (1995a,b)
Wheat, foxtail-millet, maize, rice	Moore et al. (1995a)

unadapted, as well as cultivated germplasm resources for crop improvement, two main approaches are listed below.

A. Advanced backcross QTL analysis

In order to exploit the potential of wild species in breeding programs, efforts were made in past to introduce alien or exotic genes from wild species into cultivated varieties. For example, 57 genes for resistance to diseases and pests were introduced into wheat from other genera of the *Triticeae* via alien translocations (transferring of chromosomal segments from wild or other species that carry disease resistance genes). In many cases, the size of the alien fragments and the translocation breakpoints have been precisely determined by genomic *in situ* hybridization (for review see Friebe et al., 1996). For transferring the QTLs for important traits from a wild species to a crop variety, an approach named "advanced-backcross QTL analysis (AB-QTL)" was proposed by Tanksley and Nelson (1996). In this approach, a wild species is backcrossed to a superior cultivar, and during backcrossing cycles, the transfer of desirable gene/QTL is monitored with molecular markers. The segregating BC_2F_2 or BC_2F_3 population generated during backcrossing (F_2 or F_3 stages) is then used not only for recording data on the trait of interest, but also for genotyping with polymorphic molecular markers. These data are then used for QTL analysis, leading to simultaneous discovery of QTLs. Once favorable QTL alleles are identified, only a few additional marker-assisted generations are required to develop NILs that can be field-tested and used for variety development. Therefore, a cycle of AB-QTL analysis (i.e., QTL discovery, NIL development, and testing) represents a direct test of the underlying assumption of QTL breeding: that beneficial alleles identified in segregating populations (such as BC_2 or BC_3 in the case of AB-QTL) will continue to exert their positive effects when transferred in the genetic background of elite lines (Grandillo and Tanksley, 2005).

AB-QTL analysis has been used in some studies of wheat and barley, showing that certain genomic regions (QTLs) derived from wild or unadapted germplasm have the potential to improve yield. For example, after genotyping 72 preselected BC_2F_2 plants derived from a cross between a German winter wheat variety (Prinz) and the synthetic hexaploid wheat line (W7984), Huang et al. (2003) identified a total of 40 putative QTLs involved in yield and yield component traits. For 24 (60%) of these QTLs, alleles from the synthetic wheat W7984 were associated with a favorable effect on the analyzed traits, despite the fact that synthetic wheat was overall inferior with respect to agronomic appearance and performance. For four of the seven QTLs to associate with yield, the wild-type (WT) allele had an effect that increased total yield, and the increases associated with the WT allele ranged from 5 to 15%. By using 111 BC_2F_1 lines from another cross of the wheat variety Flair with the synthetic wheat line XX86, a total of 57 QTLs were identified for seven agronomic traits analyzed (Huang et al., 2004). For 24 (42.1%) QTLs derived from XX86 line, a positive effect was observed on traits such as 1000-grain weight and number of grains per year.

In barley, the first AB-QTL study was conducted on 136 BC_2F_2 families derived from the cross of the German spring barley variety Apex and Israeli wild barley accession ISR101–23 (Pillen et al., 2003). A relatively high proportion (36%) of the 86 QTLs identified for 13 quantitative traits measured in a maximum of six environments had favorable effects derived from the exotic parent for 7 of 13 traits investigated. Interestingly, in one case, the exotic parent allele was associated with a yield increase of 7.7%, averaged across the six environments tested. To validate the QTL effects, a second AB-QTL study was undertaken by using the same wild accession (ISR101–23) but with the German spring variety Harry as the recurrent parent; 101 BC_2F_2 families were evaluated for the same 13 quantitative traits (Pillen et al., 2004). In this study, a total of 108 putative QTLs were detected, and altogether 52 (48%) favorable effects were identified from the exotic parent. The comparison of these two AB-QTL studies using the common exotic donor parent showed that in total 26% of the putative QTLs could be detected in both AB populations. Wild barley germplasm (accession HOR1508) has also proven to be a good source of QTL alleles with favorable effects on yield and other agronomically important traits under conditions of water deficit in Mediterranean countries (Forster et al., 2000; Talamè et al., 2004). Of the total 80 significant QTLs identified by Talamè et al. (2004), 42 (52%) had beneficial alleles derived from the donor wild parental line. After genotyping 181 BC_3DH lines (var Brenda × wild accession HS213), Li et al. (2005) identified a total of 25 QTLs for yield, yield components, and malting quality traits. In contrast to the previous studies, most positive QTLs originated from the recurrent parent "Brenda." One QTL each for yield and heading date (derived from Brenda) explained 18.3 and 20.7% of the phenotypic variation, respectively. This may provide a first hint that not any accession of wild germplasm will show a positive effect on a given agronomic trait.

Overall, these results have demonstrated that the AB-QTL strategy represents a very effective way to unlock valuable wild alleles and transfer them into elite cultivars to improve their performance. It is, however, important to note that in early backcross generations plants still contain a number of wild species chromosome segments which can mask the magnitude of some of favorable effects of introgressed alleles (Septiningshi et al., 2003). On the other hand, by utilizing AB-QTL approach it has been demonstrated in tomato that the pyramiding of independent yield-promoting segments introduces a number of alleles, with favorable effects, into a given genetic background after generating segmental introgression lines (ILs). These studies led to production of novel varieties that reproducibly increase productivity relative to leading commercial genotypes both under normal cultivation conditions and in the stressed environment (Gur and Zamir, 2004).

B. Association mapping based on linkage disequilibrium

Another approach toward exploiting the potential of unadapted germplasm (natural population) is utilization of the germplasm in association mapping, based on linkage disequilibrium (LD). Unlike conventional segregating (or mapping) populations such as DH, F_2, or RILs that have been used in past for identification of genes or QTLs for trait of interest for plant breeding, the natural populations are the products of many cycles of recombinations and have the potential to show enhanced resolution of QTLs. Association analysis based on LD may offer more power than linkage analysis for identifying the genes responsible for the variation in a quantitative trait (Buckler and Thornsberry, 2002; for review see Flint-Garcia et al., 2003; Gupta et al., 2005). LD is the nonrandom association of markers in a population and can provide high resolution maps of markers and genes. The extent of LD around a locus determines the resolution of association analyses and the number of markers that would be required to scan the entire genome (Rafalski and Morgante, 2004). Because genetic recombination is not evenly distributed over the genomes of most species, the linkage distance between markers and candidate genes varies widely (Philips and Vasil, 2001). Simulations estimating the power of detecting the association of variation in a candidate gene with the phenotype indicate that population size is important (Long and Langley, 1999). For a population size of 500, there is a high probability of detecting the association, even when the gene accounts for as little as 8% of the variation. For a population size of 100, only gene effects accounting for at least 15% of the variation can be detected.

LD in a germaplasm collection is affected by several factors such as recombination rate, mating system, genetic isolation, population size, population admixture, and natural and artificial selection (Rafalski and Morgante, 2004). Because of these matters, in wild species LD may extend only to a few kilobases (see Wall and Pritchard, 2003); but LD in cultivated and inbred species, such as wheat or barley, frequently extends across large linkage blocks (often almost entire chromosome arms) that have been maintained over by selection. This is particularly important in wheat where large chromosome segments from wild relatives have often been used in modern varieties, and these can show very low levels of recombination (e.g., Paull et al., 1994). However, this may also help in the localization of genes from wild relatives. For instance, Paull et al. (1998) used coefficients of parentage in an association mapping study to identify the positions of several disease resistance genes from wild relatives in wheat.

In contrast to the extensive use of LD-based association mapping in human genetics, the potential of LD-based association mapping has not been realized adequately in plant species. One of the reasons for this involves occurrence of the structured populations. In this context, Pritchard et al. (2000) proposed a population-based method that allows for large-scale assessment of allele/trait relationships in structured populations. By using this approach, association mapping based on LD has been demonstrated in maize for the Dwarf8 gene, which is involved in flowering time (Thornsberry et al., 2001) and yellow endosperm color (Palaisa et al., 2003).

In some studies, population structure has been analyzed in details in wheat as well as in barley for conducting the association mapping. For instance, after examining the levels of LD within and between 18 nuclear genes in 25 accessions from across the geographic range of wild barley, Morrell et al. (2005) demonstrated the following: (1) For the majority of wild barley loci, intralocus

LD decays rapidly, that is, at a rate similar to that observed in the out-crossing species, *Zea mays* (maize). (2) Excess interlocus LD was observed at 15% of the two-locus combinations; almost all interlocus LD involved loci that showed significant geographic structuring. However, in a collection of 134 durum wheat lines, after genotyping with 70 SSR markers, high levels of LD were found at both linked and unlinked locus pairs (Maccaferri *et al.*, 2005). Further, the information obtained from LD analysis was successfully utilized in some cases for association mapping studies. Using 236 AFLP markers in a set of 146 modern spring barley cultivars, 18–20 markers that accounted for 40–58% of the variation for yield and yield stability traits were identified (Kraakman *et al.*, 2004). Likewise in wheat, after analyzing the population structure and LD, association of 62 loci on chromosomes 2D, 5A, and 5B with kernel morphology and milling quality has been analyzed (Breseghello and Sorrells, 2005). Significant marker associations for kernel size were detected on the three chromosomes tested, and alleles potentially useful for selection were identified. This result was in agreement with previous QTL analysis.

Such high-resolution mapping of traits/QTLs to the level of individual genes will provide a new possibility for studying the molecular and biochemical basis of quantitative traits variation and will help to identify specific targets for crop improvement. It seems that association mapping approaches are viable alternative to classical QTL approaches based on crosses between inbred lines, especially for complex traits with costly measurements. However, in our opinion, though LD-based approaches hold great promise for speeding up the fine mapping, conventional linkage mapping will continue to be useful particularly when one tries to assess QTLs and the effect of a QTL in isolation (Rafalski and Morgante, 2004). In some studies, the utility of an approach involving the use of conventional linkage mapping along with LD has been recommended for the construction of molecular maps and for QTL analysis (Nordborg *et al.*, 2002; Zhu *et al.*, 2002).

VI. CONCLUDING REMARKS

Similar to other major crop species, the genetic maps of wheat and barley have benefited from the availability of large numbers of mapped markers. These have become a core resource for QTL analysis, trait-based molecular breeding, and whole-genome breeding. However, further development of existing breeding concepts will critically depend on the completion of our knowledge on genes, which underlie agronomic traits of interest. There are several ways that can be applied for the identification of candidate genes, each of which has pros and cons. Association mapping might provide a viable alternative to map-based cloning, in case there is sufficient decrease of LD around the locus of interest. At the level of functional genomics, transcript profiling may provide candidate genes for agronomic traits. However, despite the tremendous progress in structural and functional genome analysis, the rate-limiting step regarding the isolation of candidate genes will be population development and accurate phenotyping. Additional efforts are required to develop an infrastructure for phenotypic analysis of large numbers of individuals under highly standardized and reproducible conditions. Similarly, more knowledge about the physiology, cell biology, and biochemistry of the individual traits is required to break down complex traits into components that show a high heritability and can be measured accurately. Once the target genes have been identified, their genetic diversity can be studied to identify superior alleles. Notwithstanding these limitations, the availability of a comprehensive portfolio of resources for genome analysis in wheat and barley has laid the groundwork to efficiently complement existing breeding concepts and to develop knowledge-driven strategies to further adapt these cereals to our needs.

References

Aharoni, A., and Vorst, O. (2002). DNA microarrays for functional plant genomics. *Plant Mol. Biol.* **48,** 99–118.

Ahn, S., Anderson, J. A., Sorrells, M. E., and Tanksley, S. D. (1993). Homeologous relationships of rice, wheat and maize chromosomes. *Mol. Gen. Genet.* **241,** 483–490.

Andersen, J. R., and Lubberstedt, T. (2003). Functional markers in plants. *Trends Plant Sci.* **8,** 554–560.

Angerer, N., Lengauer, D., Steiner, B., Lafferty, J., Loeschenberger, F., and Buerstmayr, H. (2003). Validation of molecular markers linked to two Fusarium head blight resistance QTLs in wheat. *In* "Proceedings of the Tenth International Wheat Genetics Symposium" (N. E. Pogna, M. Romano, E. A. Pogna, and G. Galterio, eds.), pp. 1096–1098. Paestum, Italy.

Ayala, L., Henry, M., Gonzalez-de-Leon, D., Van Ginkel, M., Mujeeb-Kazi, A., Keller, B., and Khairallah, M. (2001). A diagnostic molecular marker allowing the study of *Th intermedium*-derived resistance to BYDV in bread wheat segregating populations. *Theor. Appl. Genet.* **102,** 942–949.

Ayoub, M., Armstrong, E., Bridger, G., Fortin, M. G., and Mather, D. E. (2003). Marker-based selection in barley for a QTL region affecting alpha-amylase activity of malt. *Crop Sci.* **43,** 556–561.

Blazkova, V., Bartos, P., Park, R. F., and Goyeau, H. (2002). Verifying the presence of leaf rust resistance gene *Lr10* in sixteen wheat cultivars by use of a PCR-based STS marker. *Cereal Res. Commun.* **30,** 9–16.

Bonierbale, M. W., Plaisted, R. L., and Tanksley, S. D. (1988). RFLP maps based on a common set of clones reveal modes of chromosomal evolution of potato and tomato. *Genetics* **120,** 1095–1103.

Borevitz, J. O., Liang, D., Plouffe, D., Chang, H. S., Zhu, T., Weige, D., Berry, C. C., Winzeler, E., and Chory, J. (2003). Large-scale identification of single-feature polymorphisms in complex genomes. *Genome Res.* **13,** 513–523.

Börner, A., Korzun, V., and Worland, A. J. (1998). Comparative genetic mapping of loci affecting plant height and development in cereals. *Euphytica* **100,** 245–248.

Botha, A. M., Lacock, L., Van Niekerk, C., Matsioloko, M. T., du Preez, F. B., Myburg, A. A., Kunert, K., and Cullis, C. A. (2003). Gene expression profiling during *Diuraphis noxia* infestation of *Triticum aestivum* cv 'Tugela DN' using microarrays. *In* "Proceedings of the Tenth International Wheat Genetics Symposium" (N. E. Pogna, M. Romano, E. A. Pogna, and G. Galterio, eds.), pp. 334–338. Paestum, Italy.

Brenner, S., Johnson, M., Bridgham, J., Golda, G., Lloyd, D. H., Johnson, D., Luo, S., McCurdy, S., Foy, M., Ewan, M., Roth, R., George, D., *et al.* (2000). Gene expression analysis by massively parallel signature sequencing (MPSS) on microbead arrays. *Nat. Biotechnol.* **18**, 630–634.

Breseghello, F., and Sorrells, M. E. (2005). Association mapping of kernel size and milling quality in wheat. *Genetics*, **172**, 1165–1177.

Brett, D., Pospisil, H., Valcarcel, J., Reich, J., and Bork, P. (2002). Alternative splicing and genome complexity. *Nat. Genet.* **30**, 29–30.

Buckler, E. S., and Thornsberry, J. M. (2002). Plant molecular diversity and applications to genomics. *Curr. Opin. Plant biol.* **5**, 107–111.

Cakir, M., Appels, R., Carter, M., Loughman, R., Francki, M., Li, C., Johnson, J., Bhave, M., Wilson, R., McLean, R., and Barclay, Y. (2003). Accelerated wheat breeding using molecular marker. *In* "Proceedings of the Tenth International Wheat Genetics Symposium" (N. E. Pogna, M. Romano, E. A. Pogna, and G. Galterio, eds.), pp. 117–120. Paestum, Italy.

Carter, M., Drake, B. F., Cakir, M., Jones, M., and Appels, R. (2003). Conversion of RFLP markers into PCR based markers in wheat. *In* "Proceedings of the Tenth International Wheat Genetics Symposium" (N. E. Pogna, M. Romano, E. A. Pogna, and G. Galterio, eds.), pp. 81–683. Paestum, Italy.

Close, T., Wanamaker, S. I., Caldo, R. A., Turner, S. M., Ashlock, D. A., Dickerson, J. A., Wing, R. A., Muehlbauer, G. J., Kleinhofs, A., and Wise, R. P. (2004). A new resource for cereal genomics: 22k barley genechip comes of age. *Plant Physiol.* **134**, 960–968.

Cui, X., Xu, J., Asghar, R., Condamine, P., Svensson, J. T., Wanamaker, S., Stein, N., Roose, M., and Close, T. J. (2005). Detecting single-feature polymorphisms using oligonucleotide arrays and robustified projection pursuit. *Bioinformatics* **21**, 3852–3858.

Czechowski, T., Bari, R. P., Stitt, M., Scheible, W. R., and Udvardi, M. K. (2004). Real-time RT PCR profiling of over 1400 *Arabidopsis* transcription factor genes: Unprecedented sensitivity reveal novel root- and shoot-specific genes. *Plant J.* **38**, 366–379.

Desprez, T., Amselem, J., Caboche, M., and Hofte, H. (1998). Differential gene expression in *Arabidopsis* monitored using cDNA arrays. *Plant J.* **14**, 643–652.

Devos, K. M. (2005). Updating the 'crop circle.' *Curr. Opin. Plant Biol.* **8**, 155–162.

Devos, K. M., and Gale, M. D. (1993). Extended genetic maps of the homoeologous group-3 chromosomes of wheat, rye and barley. *Theor. Appl. Genet.* **85**, 649–652.

Devos, K. M., and Gale, M. D. (1997). Comparative genetics in the grasses. *Plant Mol. Biol.* **35**, 3–15.

Devos, K. M., and Gale, M. D. (2000). Genome relationships: The grass model in current research. *Plant Cell* **12**, 637–646.

Devos, K. M., Atkinson, M. D., Chinoy, C. N., Liu, C. J., and Gale, M. D. (1992). RFLP-based genetic map of the homoeologous group 3 chromosomes of wheat and rye. *Theor. Appl. Genet.* **83**, 931–939.

Devos, K. M., Atkinson, M. D., Chinoy, C. N., Francis, H. A., Harcourt, R. L., Koebner, R. M. D., Liu, C. J., Masojć, P., Xie, D. X., and Gale, M. D. (1993a). Chromosomal rearrangement in the rye genome relative to that of wheat. *Theor. Appl. Genet.* **85**, 673–680.

Devos, K. M., Millan, T., and Gale, M. D. (1993b). Comparative RFLP maps of the homoeologous group-2 chromosomes of wheat, rye and barley. *Theor. Appl. Genet.* **85**, 784–792.

Devos, K. M., Chao, S., Li, Y., Simonetti, M. C., and Gale, M. D. (1994). Relationship between chromosome 9 of maize and wheat homoeologous group 7 chromosomes. *Genetics* **138**, 1287–1292.

Distelfeld, A., Uauy, C., Olmos, S., Schlatter, A. R., Dubcovsky, J., and Fahima, T. (2004). Microcolinearity between a 2 cM region encompassing the grain protein content locus *Gpc-6B1* on wheat chromosome 6B and a 350 kb region on rice chromosome 2. *Funct. Integr. Genomics* **4,** 59–66.

Dubcovsky, J., Luo, M. C., Zhong, G. Y., Bransteitter, R., Desai, A., Kilian, A., Kleinhofs, A., and Dvorak, J. (1996). Genetic map of diploid wheat, *Triticum monococcum* L., and its comparison with maps of *Hordeum vulgare* L. *Genetics* **143,** 983–999.

Dunford, R. P., Kurata, N., Laurie, D. A., Money, T. A., Minobe, Y., and Moore, G. (1995). Conservation of fine-scale DNA marker order in the genomes of rice and the *Triticeae*. *Nucleic Acids Res.* **23,** 2724–2728.

Ellis, M. H., Spielmeyer, W., Gale, K. R., Rebetzke, G. J., and Richards, R. A. (2002). "Perfect" markers for the *Rht-B1b* and *Rht-D1b* dwarfing genes in wheat. *Theor. Appl. Genet.* **105,** 1038–1042.

Eujayl, I., Sorrells, M. E., Baum, M., Wolters, P., and Powell, W. (2002). Isolation of EST-derived microsatellite markers for genotyping the A and B genomes of wheat. *Theor. Appl. Genet.* **104,** 399–407.

Fischer, G., Ibrahim, S. M., Brockmann, G. A., Pahnke, J., Bartocci, E., Thiesen, H. J., Serrano-Fernandez, P., and Möller, S. (2003). Expression view: Visualization of quantitative trait loci and gene-expression data in Ensembl. *Genome Biol.* **4,** R477.

Flint-Garcia, S. A., Thornsberry, J. M., and Buckler, E. S., IV. (2003). Structure of linkage disequilbrium in plants. *Annu. Rev. Plant Biol.* **54,** 357–374.

Forster, B. P., Ellis, R. P., Thomas, W. T., Newton, A. C., Tuberosa, R., This, D., El-Enein, R. A., Bahri, M. H., and Ben Salem, M. (2000). The development and application of molecular markers for abiotic stress tolerance in barley. *J. Exp. Bot.* **51,** 19–27.

Francki, M. G., Appels, R., Hunter, A., and Bellgard, M. (2003). Comparative organization of 3BS and 7AL using wheat-rice synteny. *In* "Proceedings of the Tenth International Wheat Genetics Symposium" (N. E. Pogna, M. Romano, E. A. Pogna, and G. Galterio, eds.), pp. 254–257. Paestum, Italy.

Friebe, B., Gill, K. S., Tuleen, N. A., and Gill, B. S. (1996). Transfer of wheat streak mosaic resistance from *Agropyron intermedium* into wheat. *Crop Sci.* **36,** 857–861.

Fulton, T. M., van der Hoeven, R., Eannetta, N. T., and Tanksley, S. D. (2002). Identification, analysis, and utilization of conserved ortholog set markers for comparative genomics in higher plants. *Plant Cell* **14,** 1457–1467.

Gallego, F., Feuillet, C., Messmer, M., Penger, A., Graner, A., Yano, M., Sasaki, T., and Keller, B. (1998). Comparative mapping of the two wheat leaf rust resistance loci *Lr1* and *Lr10* in rice and barley. *Genome* **41,** 328–336.

Gao, L. F., Jing, R. L., Huo, N. X., Li, Y., Li, X. P., Zhou, R. H., Chang, X. P., Tang, J. F., Ma, Z. W., and Jia, J. Z. (2004). One hundred and one new microsatellite loci derived from ESTs (EST-SSRs) in bread wheat. *Theor. Appl. Genet.* **108,** 1392–1400.

Grandillo, S., and Tanksley, S. D. (2005). Advanced backcross QTL analysis: Results and perspectives. *In* "In the Wake of Double Helix: From the Green Revolution to the Gene Revolution" (R. Tuberosa, R. L. Phillips, and M. Gale, eds.), pp. 115–132. Avenue Media, Bologna, Italy.

Graner, A., Thiel, T., Zhang, H., Potokina, E., Prasad, M., Perovic, D., Kota, R., Varshney, R. K., Scholz, U., Grosse, I., and Stein, N. (2005). Molecular mapping in barley: Shifting from the structural to the functional level. *Czech J. Genet. Plant Breed.* **41,** 81–88.

Gudu, S., Laurie, D. A., Kasha, K. J., Xia, J. J., and Snape, J. W. (2002). RFLP mapping of a *Hordeum bulbosum* gene highly expressed in pistils and its relationship to homoeologous loci in other Gramineae species. *Theor. Appl. Genet.* **105,** 271–276.

Gupta, P. K., and Varshney, R. K. (2000). The development and use of microsatellite markers for genetic analysis and plant breeding with emphasis on bread wheat. *Euphytica* **113,** 163–185.

Gupta, P. K., and Varshney, R. K. (2004). Cereal genomics: An overview. *In* "Cereal Genomics" (P. K. Gupta and R. K. Varshney, eds.), pp. 1–18. Kluwer Academic Publishers, Dordrecht, The Netherlands.

Gupta, P. K., Varshney, R. K., Sharma, P. C., and Ramesh, B. (1999). Molecular markers and their applications in wheat breeding. *Plant Breed.* **118,** 369–390.

Gupta, P. K., Balyan, H. S., Edwards, K. J., Isaac, P., Korzun, V., Röder, M., Gautier, M. F., Joudrier, P., Schlatter, A. R., Dubcovsky, J., De la Pena, R., Khairallah, M., *et al.* (2002a). Genetic mapping of 66 new microsatellite (SSR) loci in bread wheat. *Theor. Appl. Genet.* **105,** 413–422.

Gupta, P. K., Varshney, R. K., and Prasad, M. (2002b). Molecular markers: Principles and methodology. *In* "Molecular Techniques in Crop Improvement" (S. M. Jain, D. S. Brar, and B. S. Ahloowalia, eds.), pp. 9–54. Kluwer Academic Publishers, Dordrecht, The Netherlands.

Gupta, P. K., Kulwal, P., and Rustgi, S. (2005). Linkage disequilibrium and association studies in higher plants: Present status and future prospects. *Plant Mol. Biol.* **57,** 461–485.

Gur, A., and Zamir, D. (2004). Unused natural variation can lift yield barriers in plant breeding. *PLoS Biol.* **2,** e245.

Guyomarc'h, H., Sourdille, P., Edwards, K. J., and Bernard, M. (2002). Studies of the transferability of microsatellites derived from *Triticum tauschii* to hexaploid wheat and to diploid related species using amplification, hybridization and sequence comparisons. *Theor. Appl. Genet.* **105,** 736–744.

Guyot, R., Yahiaoui, N., Feuillet, C., and Keller, B. (2004). *In silico* comparative analysis reveals a mosaic conservation of genes within a novel colinear region in wheat chromosome 1AS and rice chromosome 5S. *Funct. Integr. Genomics* **4,** 47–58.

Han, F., Romagosa, I., Ullrich, S. E., Jones, B. L., Hayes, P. M., and Wesenberg, D. M. (1997). Molecular marker-assisted selection for malting quality traits in barley. *Mol. Breed.* **3,** 427–437.

Han, F., Kleinhofs, A., Ullrich, S. E., Kilian, A., Yano, M., and Sasaki, T. (1998). Synteny with rice: Analysis of barley malting quality QTLs and *rpg4* chromosome regions. *Genome* **41,** 373–380.

Han, F., Kilian, A., Chen, J. P., Kudrna, D., Steffenson, B., Yamamoto, K., Matsumoto, T., Sasaki, T., and Kleinhofs, A. (1999). Sequence analysis of a rice BAC covering the syntenous barley *Rpg1* region. *Genome* **42,** 1071–1076.

Hayes, P. M., Corey, A. E., Mundt, C., Toojinda, T., and Vivar, H. (2003a). Registration of 'Tango' barley. *Crop Sci.* **43,** 729–731.

Hayes, P. M., Castro, A., Marquez-Cedillo, L., Corey, A., Henson, C., Jones, B. L., Kling, J., Mather, D., Matus, I., Rossi, C., *et al.* (2003b). Genetic diversity for quantitatively inherited agronomic and malting quality traits. *In* "Diversity in Barley (*Hordeum vulagre*)" (R. von Bothmer, Th. Van Hintum, H. Knüpffer, and K. Sato, eds.), pp. 201–226. Elsevier Science B.V., Amsterdam, The Netherlands.

Hernandez, P., Dorado, G., Prieto, P., Gimenez, M. J., Ramirez, M. C., Laurie, D. A., Snape, J. W., and Martin, A. (2001). A core genetic map of *Hordeum chilense* and comparisons with maps of barley (*Hordeum vulgare*) and wheat (*Triticum aestivum*). *Theor. Appl. Genet.* **102,** 1259–1264.

Hohmann, U., Graner, A., Endo, T. R., Gill, B. S., and Herrmann, R. G. (1995). Comparison of wheat physical maps with barley linkage maps for group 7 chromosomes. *Theor. Appl. Genet.* **91,** 618–626.

Horak, C. E., and Snyder, M. (2002). Global analysis of gene expression in yeast. *Funct. Integr. Genomics* **2,** 171–180.

Huang, X. Q., Cöster, H., Ganal, M. W., and Röder, M. S. (2003). Advanced backcross QTL analysis for the identification of quantitative trait loci alleles from wild relatives of wheat (*Triticum aestivum* L.). *Theor. Appl. Genet.* **106,** 1379–1389.

Huang, X. Q., Cöster, H., Kemp, H., Ganal, M. W., and Röder, M. S. (2004). Advanced backcross QTL analysis in progenies derived from a cross between a German elite winter wheat variety and a synthetic wheat (*Triticum aestivum* L.). *Theor. Appl. Genet.* **109,** 933–943.

Jahoor, A., Eriksen, L., and Backes, G. (2004). QTLs and genes for disease resistance in barley and wheat. *In* "Cereal Genomics" (P. K. Gupta and R. K. Varshney, eds.), pp. 199–252. Kluwer Academic Publishers, Dordrecht, The Netherlands.

Jansen, R. C. (2003). Studying complex biological systems using multifactorial perturbation. *Nat. Rev. Genet.* **4**, 145–151.

Jansen, R. C., and Nap, J. P. (2001). Genetical genomics: The added value from segregation. *Trends Genet.* **17**, 388–391.

Jefferies, S. P. (2000). Marker-assisted backcrossing for gene introgression in barley (*Hordeum vulgare* L.). Ph.D. Thesis, University of Adelaide, Australia.

Jefferies, S. P., King, B. J., Barr, A. R., Warner, P., Logue, S. J., and Langridge, P. (2003). Marker-assisted backcross introgression of the yd2 gene conferring resistance to barley yellow dwarf virus in barley. *Plant Breed.* **122**, 52–56.

Johnson, J. M., Castle, J., Garrett-Engele, P., Kan, Z., Loerch, P. M., Armour, C. D., Santos, R., Schadt, E. E., Stoughton, R., and Shoemaker, D. D. (2003). Genome-wide survey of human alternative pre-mRNA splicing with exon junction microarrays. *Science* **302**, 2141–2144.

Kato, K., Miura, H., and Sawada, S. (1999). Comparative mapping of the wheat Vrn-A1 region with the rice Hd-6 region. *Genome* **42**, 204–209.

Kato, K., Nakamura, W., Tabiki, T., Miura, H., and Sawada, S. (2001). Detection of loci controlling seed dormancy on group 4 chromosomes of wheat and comparative mapping with rice and barley genomes. *Theor. Appl. Genet.* **102**, 980–985.

Khlestkina, E. K., Than, M. H. M., Pestsova, E. G., Röder, M. S., Malyshev, S. V., Korzun, V., and Börner, A. (2004). Mapping of 99 microsatellite loci in rye (*Secale cereale* L.) including 39 expressed sequence tags. *Theor. Appl. Genet.* **109**, 725–732.

Kilian, A., Kudrna, D. A., Kleinhofs, A., Yano, M., Kurata, N., Steffenson, B., and Sasaki, T. (1995). Rice-barley synteny and its application to saturation mapping of the barley Rpg1 region. *Nucleic Acids Res.* **23**, 2729–2733.

Kilian, A., Chen, J., Han, F., Steffenson, B., and Kleinhofs, A. (1997). Towards map-based cloning of the barley stem rust resistance genes Rpg1 and rpg4 using rice as an intergenomic cloning vehicle. *Plant Mol. Biol.* **35**, 187–195.

Koebner, R. M. D., Powell, W., and Donini, R. M. D. (2001). Contributions of DNA molecular marker technologies to the genetics and breeding of wheat and barley. *Plant Breed. Rev.* **21**, 181–220.

Kota, R., Varshney, R. K., Thiel, T., Dehmer, K. J., and Graner, A. (2001). Generation and comparison of EST-derived SSR and SNP markers in barley (*Hordeum vulgare* L.). *Hereditas* **135**, 141–151.

Kota, R., Rudd, S., Facius, A., Kolesov, G., Thiel, T., Zhang, H., Stein, N., Mayer, K., and Graner, A. (2003). Snipping polymorphisms from large EST collections in barley (*Hordeum vulgare* L.). *Mol. Gen. Genomics* **270**, 24–33.

Kraakman, A. T. W., Niks, R. E., van den Berg, P. M. M. M., Stam, P., and van Eeuwijk, F. A. (2004). Linkage disequilibrium mapping of yield and yield stability in modern spring barley cultivars. *Genetics* **168**, 435–446.

Kuchel, H., Wraner, P., Fox, R. L., Chalmers, K., Howes, N., Langridge, P., and Jefferies, S. P. (2003). Whole genome based marker assisted selection strategies in wheat breeding. *In* "Proceedings of the Tenth International Wheat Genetics Symposium" (N. E. Pogna, M. Romano, E. A. Pogna, and G. Galterio, eds.), pp. 144–147. Paestum, Italy.

Kuchel, H., Ye., G., Fox, R., and Jefferies, S. (2005). Genetic and economic analysis of a targeted marker-assisted wheat breeding strategy. *Mol. Breed.* **16**, 67–78.

Kurata, N., Moore, G., Nagamura, Y., Foote, T., Yano, M., Minobe, Y., and Gale, M. (1994). Conservation of genomic structure between rice and wheat. *Bio/Technology* **12**, 276–278.

La Rota, C. M., and Sorrells, M. E. (2004). Comparative DNA sequence analysis of mapped wheat ESTs reveals complexity of genome relationships between rice and wheat. *Funct. Integr. Genomics* **4,** 34–46.

Lagercrantz, U. (1998). Comparative mapping between *Arabidopsis thaliana* and *Brassica nigra* indicates that *Brassica* genomes have evolved through extensive genome replication accompanied by chromosome fusions and frequent rearrangements. *Genetics* **150,** 1217–1228.

Lamoureux, D., Boeuf, C., Regad, F., Garsmeur, O., Charmet, G., Sourdille, P., Lagoda, P., and Bernard, M. (2002). Comparative mapping of the wheat 5B short chromosome arm distal region with rice, relative to a crossability locus. *Theor. Appl. Genet.* **105,** 759–765.

Langridge, P. (2005). Molecular breeding of wheat and barley. *In* "In the Wake of Double Helix: From the Green Revolution to the Gene Revolution" (R. Tuberosa, R. L. Phillips, and M. Gale, eds.), pp. 279–286. Avenue Media, Bologna Italy.

Langridge, P., and Chalmers, K. (1998). Techniques for marker development. *In* "Proceedings of the Ninth International Wheat Genetics Symposium" (A. E. Slinkard, ed.), Vol. 1, pp. 107–117. University Extension Press, University of Saskatchewan, Saskatoon, Canada.

Langridge, P., and Chalmers, K. (2004). The principle: Identification and application of molecular markers. *In* "Biotechnology in Agriculture and Forestry, Vol 55, Molecular Markers Systems" (H. Lörz and G. Wenzel, eds.), pp. 3–22. Springer Verlag, Heidelberg, Germany.

Langridge, P., Lagudah, E. S., Holton, T. A., Appels, R., Sharp, P. J., and Chalmers, K. J. (2001). Trends in genetic and genome analyses in wheat: A review. *Aust. J. Agric. Res.* **52,** 1043–1077.

Larkin, J. E., Frank, B. C., Gavras, H., Sultana, R., and Quackenbush, J. (2005). Independence and reproducibility across microarray platforms. *Nat. Methods* **2,** 337–343.

Laubin, B., Nicot, N., Amiour, N., Sourdille, P., Branland, G., and Leroy, P. (2003). *In silico* mapping and colinearity between homoeologous group 5 of wheat and the rice genome. *In* "Proceedings of the Tenth International Wheat Genetics Symposium" (N. E. Pogna, M. Romano, E. A. Pogna, and G. Galterio, eds.), pp. 280–283. Paestum, Italy.

Lazo, G. R., Chao, S., Hummel, D., Edwards, H., Crosman, C. C., Lui, N., Matthews, D. E., Carollo, V. L., Hane, D. L., You, F. M., Butler, G. E., Miller, R. E., *et al.* (2004). Development of an expressed sequence tag (EST) resource for wheat (*Triticum aestivum*): EST generation, unigene analysis, probe selection and bioinformatics for a 16,000 locus bin-delineated map. *Genetics* **168,** 585–593.

Leader, D. J., Cullup, T., Ridley, P., and van Dodeweerd, A. M. (2003). Microarray analysis of wheat grain development: Applications to trait charcterization in the field. *In* "Proceedings of the Tenth International Wheat Genetics Symposium" (N. E. Pogna, M. Romano, E. A. Pogna, and G. Galterio, eds.), pp. 287–292. Paestum, Italy.

Li, J. Z., Sjakste, T. G., Röder, M. S., and Ganal, M. W. (2003). Development and genetic mapping of 127 new microsatellite markers in barley. *Theor. Appl. Genet.* **107,** 1021–1027.

Li, J. Z., Huang, X. Q., Heinrichs, F., Ganal, M. W., and Röder, M. S. (2005). Analysis of QTLs for yield, yield components, and malting quality in a BC3-DH population of spring barley. *Theor. Appl. Genet.* **110,** 356–363.

Li, Z., Huang, N., Rampling, L., Wang, J., Yu, J., Morell, M., and Rahman, S. (2004). Detailed comparison between the wheat chromosome group 7 short arms and the rice chromosome arms 6S and 8L with special reference to genes involved in starch biosynthesis. *Funct. Integr. Genomics* **4,** 231–240.

Liu, Z. W., Biyashev, R. M., and Saghai-Maroof, M. A. (1996). Development of simple sequence repeat DNA markers and their integration into a barley linkage map. *Theor. Appl. Genet.* **93,** 869–876.

Liu, J., Liu, D., Tao, W., Li, W., Wang, S., Chen, P., Cheng, S., and Gao, D. (2000). Molecular marker-facilitated pyramiding of different genes for powdery mildew resistance in wheat. *Plant Breed.* **119,** 21–24.

Liu, S. X., and Anderson, J. A. (2003). Targeted molecular mapping of a major wheat QTL for *Fusarium* head blight resistance using wheat ESTs and synteny with rice. *Genome* **46**, 817–823.

Long, A. D., and Langley, C. H. (1999). The power of association studies to detect the contribution of candidate genetic loci to variation in complex traits. *Genome Res.* **9**, 720–731.

Maccaferri, M., Sanguineti, M., Noli, E., and Tuberosa, R. (2005). Population structure and long-range linkage disequilibrium in a durum wheat elite collection. *Mol. Breed.* **15**, 271–290.

Mares, D. J., and Mrva, K. (2001). Mapping quantitative trait loci associated with variation in grain dormancy in Australian wheat. *Aust. J. Agric. Res.* **52**, 1257–1265.

Matsumura, H., Reich, S., Akiko, I., Hiromasa, S., Sophien, K., Peter, W, Günter, K., Monika, R., Detlev, R., Krüger, H., and Terauchi, R. (2003). Gene expression analysis of plant host–pathogen interactions by SuperSAGE. *Proc. Natl. Acad. Sci. USA* **100**, 15718–15723.

Miklos, G. L., and Maleszka, R. (2004). Microarray reality checks in the context of a complex disease. *Nat. Biotechnol.* **22**, 615–621.

Milligan, A. S., Lopato, S., and Langridge, P. (2004). Functional genomics studies of seed development in cereals. *In* "Cereal Genomics" (P. K. Gupta and R. K. Varshney, eds.), pp. 447–482. Kluwer Academic Publishers, Dordrecht, The Netherlands.

Moore, G., Devos, K. M., Wang, Z., and Gale, M. D. (1995a). Cereal genome evolution—grasses, line up and form a circle. *Curr. Biol.* **5**, 737–739.

Moore, G., Foote, T., Helentjaris, T., Devos, K. M., Kurata, N., and Gale, M. D. (1995b). Was there a single ancestral cereal chromosome? *Trends Genet.* **11**, 81–82.

Morrell, P. L., Toleno, D. M., Lundy, K. E., and Clegg, M. T. (2005). Low levels of linkage disequilibrium in wild barley (*Hordeum vulgare* ssp. *Spontaneum*) despite high rates of self-fertilization. *Proc. Natl. Acad. Sci. USA* **102**, 2442–2447.

Namuth, D. M., Lapitan, N. L. V., Gill, K. S., and Gill, B. S. (1994). Comparative RFLP mapping of *Hordeum vulgare* and *Triticum tauschii*. *Theor. Appl. Genet.* **89**, 865–872.

Nicot, N., Chiquet, V., Gandon, B., Amilhat, L., Legeai, F., Leroy, P., Bernard, M., and Sourdille, P. (2004). Study of simple sequence repeat (SSR) markers from wheat expressed sequence tags (ESTs). *Theor. Appl. Genet.* **109**, 800–805.

Nordborg, M., Borevitz, J. O., Bergelson, J., Berry, C. C., Chory, J., Hagenblad, J., Kreitman, M., Maloof, J. N., Noyes, T., Oefner, P. J., Stahl, E. A., and Weigel, D. (2002). The extent of linkage disequilibrium in *Arabidopsis thaliana*. *Nat. Genet.* **30**, 190–193.

Ogihara, Y., Mochida, K., Nemoto, Y., Murai, K., Yamazaki, Y., Shin-I, T., and Kohara, Y. (2003). Correlated clustering and virtual display of gene expression patterns in the wheat life cycle by large-scale statistical analyses expressed sequence tags. *Plant J.* **33**, 1001–1011.

Palaisa, K., Morgante, M., Williams, M., and Rafalski, A. (2003). Contrasting effects of selection on sequence diversity and linkage disequilibrium at two phytoene synthase loci. *Plant Cell* **15**, 1795–1806.

Pasquer, F., Stein, N., Isidore, E., and Keller, B. (2003). Microarray analysis of gene expression in wheat (*Triticum aestivum*) after fungicide application. *In* "Proceedings of the Tenth International Wheat Genetics Symposium" (N. E. Pogna, M. Romano, E. A. Pogna, and G. Galterio, eds.), pp. 1029–1031. Paestum, Italy.

Paterson, A. H. (2004). Comparative genomics in cereals. *In* "Cereal Genomics" (P. K. Gupta and R. K. Varshney, eds.), pp. 119–134. Kluwer Academic Publishers, Dordrecht, The Netherlands.

Paull, J. G., Pallotta., M. A., and Langridge, P. (1994). The RFLP markers associated with *Sr22* and recombination between chromosome 7A of bread wheat and the diploid species *Triticum boeoticum*. *Theor. Appl. Genet.* **89**, 1039–1045.

Paull, J. G., Chalmers, K. J., Karakousis, A., Kretschmer, J. M., Manning, S., and Langridge, P. (1998). Genetic diversity in Australian wheat varieties and breeding material based on RFLP data. *Theor. Appl. Genet.* **96**, 435–446.

Pellio, B., Streng, S., Bauer, E., Stein, N., Perovic, D., Schiemann, A., Friedt, W., Ordon, F., and Graner, A. (2005). High-resolution mapping of the Rym4/Rym5 locus conferring resistance to the barley yellow mosaic virus complex (BaMMV, BaYMV, BaYMV-2) in barley (*Hordeum vulgare* ssp. *vulgare* L.). *Theor. Appl. Genet.* **10**, 283–293.

Peng, J. H., and Lapitan, N. L. V. (2005). Characterization of EST-derived microsatellites in the wheat genome and development of eSSR markers. *Funct. Integr. Genomics* **5**, 80–96.

Perovic, D., Stein, N., Zhang, H., Drescher, A., Prasad, M., Kota, R., Kopahnke, D., and Graner, A. (2004). An integrated approach for comparative mapping in rice and barley based on genomic resources reveals a large number of syntenic markers but no candidate gene for the *Rph16* resistance locus. *Funct. Integr. Genomics* **4**, 74–83.

Pestsova, E., Ganal, M. W., and Röder, M. S. (2000). Isolation and mapping of microsatellite markers specific for the D genome of bread wheat. *Genome* **43**, 689–697.

Philips, R. L., and Vasil, I. K. (2001). "DNA-Based Markers in Plants." Kluwer Academic Publishers, The Netherlands.

Pillen, K., Binder, A., Kreuzkam, B., Ramsay, L., Waugh, R., Förster, J., and Leon, J. (2000). Mapping new EMBL-derived barley microsatellites and their use in differentiating German barley cultivars. *Theor. Appl. Genet.* **101**, 652–660.

Pillen, K., Zacharias, A., and Léon, J. (2003). Advanced backcross QTL analysis in barley (*Hordeum vulgare* L.). *Theor. Appl. Genet.* **107**, 340–352.

Pillen, K., Zacharias, A., and Leon, J. (2004). Comparative AB-QTL analysis in barley using a single exotic donor of *Hordeum vulgare* ssp. *spontaneum*. *Theor. Appl. Genet.* **108**, 1591–1601.

Potokina, E., Caspers, M., Prasad, M., Kota, R., Zhang, H., Sreenivasulu, N., Wang, M., and Graner, A. (2004). Functional association between malting quality trait components and cDNA array based expression patterns in barley (*Hordeum vulgare* L.). *Mol. Breed.* **14**, 153–170.

Potokina, E., Prasad, M., Malysheva, L., Röder, M. S., and Graner, A. (2006). Expression genetics and haplotype analysis reveal *cis* regulation of serine carboxypeptidase I (*Cxp1*), a candidate gene for malting quality in barley (*Hordeum vulgare* L.). *Funct. Integr. Genomics* **6**, 25–35.

Powell, W., and Langridge, P. (2004). Unfashionable crop species flourish in the 21st century. *Genome Biol.* **5,Art 233.**.

Pritchard, J. K., Stephens, M., Rosenberg, N. A., and Donnelly, P. (2000). Association mapping in structured populations. *Am. J. Hum. Genet.* **67**, 170–181.

Pumphrey, M. O., and Anderson, J. A. (2003). QTL validation via systematic development of near-isogenic wheat lines from existing breeding populations. *In* "Proceedings of the Tenth International Wheat Genetics Symposium" (N. E. Pogna, M. Romano, E. A. Pogna, and G. Galterio, eds.), pp. 1227–1229. Paestum, Italy.

Qi, L.-L., Echalier, B., Chao, S., Lazo, G. R., Butler, G. E., Anderson, O. D., Akhunov, E. D., Dvorak, J., Linkiewicz, A. M., Ratnasiri, A., Dubcovsky, J., Bermudez-Kandianis, C. E., *et al.* (2004). A chromosome bin map of 16,000 expressed sequence tag loci and distribution of genes among the three genomes of polyploid wheat. *Genetics* **168**, 701–712.

Rafalski, J. A. (2002). Applications of single nucleotide polymorphisms in crop genetics. *Curr. Opin. Plant Biol.* **5**, 94–100.

Rafalski, A., and Morgante, M. (2004). Corn and humans: Recombination and linkage disequilibrium in two genomes of similar size. *Trends Genet.* **20**, 103–111.

Ramsay, L., Macaulay, M., Ivanissevich, D. S., MacLean, K., Cardle, L., Fuller, J., Edwards, K. J., Tuvesson, S., Morgante, M., Massari, A., Maestri, E., Marmiroli, N., *et al.* (2000). A simple sequence repeat-based linkage map of barley. *Genetics* **156**, 1997–2005.

Röder, M. S., Korzun, V., Wendehake, K., Plaschke, J., Tixier, M., Leroy, P., and Ganal, M. W. (1998). A microsatellite map of wheat. *Genetics* **149**, 2007–2023.

Rostoks, N., Borevitz, J. O., Headley, P. E., Russel, J., Sharon, M., Jenny, M., Cardle, L., Marshall, D. F., and Robbie, W. (2005). Single-feature polymorphism discovery in the barley transcriptome. *Genome Biol.* **6,** R54.

Rudd, S., Schoof, H., and Mayer, K. (2005). Plant Markers—a database of predicted molecular markers from plants. *Nucleic Acids Res.* **33,** 628–632.

Saghai-Maroof, M. A., Tang, G. P., Biyashev, R. M., Maughan, P. J., and Zhang, Q. (1996). Analysis of the barley and rice genomes by comparative RFLP linkage mapping. *Theor. Appl. Genet.* **92,** 541–551.

Salvi, S., and Tuberosa, R. (2005). To clone or not to clone plant QTLs: Present and future challenges. *Trends Plant Sci.* **10,** 297–304.

Salvo-Garrido, H., Laurie, D. A., Jaffe, B., and Snape, J. W. (2001). An RFLP map of diploid *Hordeum bulbosum* L. and comparison with maps of barley (*H. vulgare* L.) and wheat (*Triticum aestivum* L). *Theor. Appl. Genet.* **103,** 869–880.

Sarma, R. N., Gill, B. S., Sasaki, T., Galiba, G., Sutka, J., Laurie, D. A., and Snape, J. W. (1998). Comparative mapping of the wheat chromosome 5A *Vrn-A1* region with rice and its relationship to QTL for flowering time. *Theor. Appl. Genet.* **97,** 103–109.

Sarma, R. N., Fish, L., Gill, B. S., and Snape, J. W. (2000). Physical characterization of the homoeologous Group 5 chromosomes of wheat in terms of rice linkage blocks, and physical mapping of some important genes. *Genome* **43,** 191–198.

Schadt, E. E., Monks, S. A., Drake, T. A., Lusis, A. J., Che, N., Colinayo, V., Ruff, T. G., Milligan, S. B., Lamb, J. R., Cavet, G., Linsley, P. S., Mao, M., *et al.* (2003). Genetics of gene expression surveyed in maize, mouse and man. *Nature* **422,** 297–301.

Schena, M., Shalon, D., Davis, R. W., and Brown, P. O. (1995). Quantitative monitoring of gene expression patterns with a complementary DNA microarray. *Science* **270,** 467–470.

Schulman, A. H., Gupta, P. K., and Varshney, R. K. (2004). Organization of microsatellites and retrotransposons in cereal genomes. *In* "Cereal Genomics" (P. K. Gupta and R. K. Varshney, eds.), pp. 83–118. Kluwer Academic Publishers, Dordrecht, The Netherlands.

Septiningshi, E. M., Prasetiyono, J., Lubis, E., Tai, T. H., Tjubaryat, T., Moeljopawiro, S., and McCouch, S. R. (2003). Identification of quantitative trait loci for yield and yield components in an advanced backcross population derived from the *Oryza sativa* variety IR64 and the wild relative *O. rufipogon. Theor. Appl. Genet.* **107,** 1419–1432.

Shan, X., Blake, T. K., and Talbert, L. E. (1999). Conversion of AFLP markers to sequence-specific PCR markers in barley and wheat. *Theor. Appl. Genet.* **98,** 1072–1078.

Shariflou, M. R., Ghannadha, M. R., and Sharp, P. J. (2003). Multiplex PCR of microsatellite markers in wheat. *In* "Proceedings of the Tenth International Wheat Genetics Symposium" (N. E. Pogna, M. Romano, E. A. Pogna, and G. Galterio, eds.), pp. 1050–1052. Paestum, Italy.

Shinbata, T., Vrinten, P., Iida, J., Sato, M., Yonemaru, J., Saito, M., Mitsuse, S., and Nakamura, T. (2003). Microarray analysis of gene expression in developing endosperm from different wheat varities. *In* "Proceedings of the Tenth International Wheat Genetics Symposium" (N. E. Pogna, M. Romano, E. A. Pogna, and G. Galterio, eds.), pp. 1053–1055. Paestum, Italy.

Singh, H., Prasad, M., Varshney, R. K., Roy, J. K., Balyan, H. S., Dhaliwal, H. S., and Gupta, P. K. (2001). STMS markers for grain protein content and their validation using near isogenic lines in bread wheat. *Plant Breed.* **120,** 273–278.

Singh, N. K., Raghuvanshi, S., Srivastava, S. K., Gaur, A., Pal, K., Dalal, V., Singh, A., Ghazil, I. A., Bhargav, A., Yadav, M., Dixit, A., Batra, K., *et al.* (2004). Sequence analysis of the long arm of rice chromosome 11 for rice–wheat synteny. *Funct. Integr. Genomics* **4,** 102–117.

Smilde, D. W., Haluskova, J., Sasaki, T., and Graner, A. (2001). New evidence for the synteny of rice chromosome 1 and barley chromosome 3H from rice expressed sequence tags. *Genome* **44,** 361–367.

Somers, D., Edwards, K. J., and Issac, P. (2004). A high density microsatellite consensus map for bread wheat (*Triticum aestivum* L.). *Theor. Appl. Genet.* **109**, 1105–1114.

Somers, D. J., Kirkpatrick, R., Moniwa, M., and Walsh, A. (2003). Mining single nucleotide polymorphisms from hexaploid wheat ESTs. *Genome* **49**, 431–437.

Song, Q. J., Fickus, E. W., and Cregan, P. B. (2002). Characterization of trinucleotide SSR motifs in wheat. *Theor. Appl. Genet.* **104**, 286–293.

Sorrells, M. E., La Rota, M., Bermudez-Kandianis, C. E., Greene, R. A., Kantety, R., Munkvold, J. D., Miftahudin Mahmoud, A., Ma, X., Gustafson., P. J., Qi, L. L., Echalier, B., et al. (2003). Comparative DNA sequence analysis of wheat and rice genomes. *Genome Res.* **13**, 1818–1827.

Sreenivasulu, N., Kavikishor, P. B., Varshney, R. K., and Altschmied, L. (2002). Mining functional information from cereal genomes—the utility of expressed sequence tags (ESTs). *Curr. Sci.* **83**, 965–973.

Sreenivasulu, N., Varshney, R. K., Kavikishore, P. V., and Weschke, W. (2004). Tolerance to abioitic stress—a functional genomics approach. *In* "Cereal Genomics" (P. K. Gupta and R. K. Varshney, eds.), pp. 483–514. Kluwer Academic Publishers, The Netherlands.

Stein, N., and Graner, A. (2004). Map-based gene isolation in cereal genomes. *In* "Cereal Genomics" (P. K. Gupta and R. K. Varshney, eds.), pp. 331–360. Kluwer Academic Publishers, Dordrecht, The Netherlands.

Stein, N., Feuillet, C., Wicker, T., Schlagenhauf, E., and Keller, B. (2000). Subgenome chromosome walking in wheat: A 450-kb physical contig in *Triticum monococcum* L spans the *Lr10* resistance locus in hexaploid wheat (*Triticum aestivum* L). *Proc. Natl. Acad. Sci. USA* **97**, 13436–13441.

Stephenson, P., Bryan, G., Kirby, J., Collins, A., Devos, K. M., Busso, C., and Gale, M. D. (1998). Fifty new microsatellite loci for the wheat genetic map. *Theor. Appl. Genet.* **97**, 946–949.

Talamè, V., Sanguineti, M. C., Chiapparino, E., Bahri, H., Salem, B. M., Forster, B. P., Ellis, R. P., Rhouma, S., Zoumarou, W., Waugh, R., and Tuberosa, R. (2004). Identification of *Hordeum spontaneum* QTL alleles improving field performance of barley grown under rainfed conditions. *Ann. Appl. Biol.* **144**, 309–319.

Tan, P. K., Downey, T. J., Spitznagel, E. L., Jr., Xu, P., Fu, D., Dimitrov, D. S., Lempicki, R. A., Raaka, B. M., and Cam, M. C. (2003). Evaluation of gene expression measurements from commercial microarray platforms. *Nucleic Acids Res.* **31**, 5676–5684.

Tanksley, S. D., and McCouch, S. (1997). Seed banks and molecular maps: Unlocking genetic potential from the wild. *Science.* **277**, 1063–1066.

Tanksley, S. D., and Nelson, J. C. (1996). Advanced backcross QTL analysis: A method for the simultaneous discovery and transfer of valuable QTLs from unadapted germplasm into elite breeding lines. *Theor. Appl. Genet.* **92**, 191–203.

Tanksley, S. D., Ganal, M. W., Prince, J. P., de Vicente, M. C., Bonierbale, M. W., Broun, P., Fulton, T. M., Giovannoni, J. J., Grandillo, S., Martin, G. B., Messeguer, R., Miller, J. C., et al. (1992). High density molecular linkage maps of the tomato and potato genomes. *Genetics* **132**, 1141–1160.

Thiel, T., Michalek, W., Varshney, R. K., and Graner, A. (2003). Exploiting EST databases for the developement and characterization of gene-derived SSR-markers in barley (*Hordeum vulgare* L.). *Theor. Appl. Genet.* **106**, 411–422.

Thornsberry, J. M., Goodman, M. M., Doebley, J., Kresovich, S., Nielsen, D., and Buckler, E. S. (2001). *Dwarf8* polymorphisms associate with variation in flowering time. *Nat. Genet.* **28**, 286–289.

Toojinda, T., Baird, E., Booth, A., Broers, L., Hayes, P., Powell, W., Thomas, W., Vivar, H., and Young, G. (1998). Introgression of quantitative trait loci (QTLs) determining stripe rust resistance in barley: An example of marker-assisted line development. *Theor. Appl. Genet.* **96**, 123–131.

Tuberosa, R., and Salvi, S. (2004). QTLs and genes for tolerance to abiotic stress in cereals. *In* "Cereal Genomics" (P. K. Gupta and R. K. Varshney, eds.), pp. 253–315. Kluwer Academic Publishers, Dordrecht, The Netherlands.

Van Deynze, A. E., Dubcovsky, J., Gill, K. S., Nelson, J. C., Sorrells, M. E., Dvorak, J., Gill, B. S., Lagudah, E. S., McCouch, S. R., and Appels, R. (1995a). Molecular-genetic maps for group 1 chromosomes of *Triticeae* species and their relation to chromosomes in rice and oat. *Genome* **38**, 45–59.

Van Deynze, A. E., Nelson, J. C., Yglesias, E. S., Harrington, S. E., Braga, D. P., McCouch, S. R., and Sorrells, M. E. (1995b). Comparative mapping in gasses–wheat relationships. *Mol. Gen. Genet.* **248**, 744–754.

Varshney, R. K., Thiel, T., Stein, N., Langridge, P., and Graner, A. (2002). *In silico* analysis on frequency and distribution of microsatellites in ESTs of some cereal species. *Cell. Mol. Biol. Lett.* **7**, 537–546.

Varshney, R. K., Korzun, V., and Börner, A. (2004a). Molecular maps in cereals: Methodology and progress. *In* "Cereal Genomics" (P. K. Gupta and R. K. Varshney, eds.), pp. 35–82. Kluwer Academic Publishers, Dordrecht, The Netherlands.

Varshney, R. K., Prasad, M., and Graner, A. (2004b). A Molecular marker maps of barley: A resource for intra- and interspecific genomics. *In* "Biotechnology in Agriculture and Forestry, Vol. 55, Molecular Markers Systems" (H. Lörz and G. Wenzel, eds.), pp. 229–245. Springer Verlag, Heidelberg, Germany.

Varshney, R. K., Zhang, H., Potokina, E., Stein, N., Langridge, P., and Graner, A. (2004c). A simple hybridisation-based strategy for the generation of non-redundant EST collections. *Plant Sci.* **167**, 629–634.

Varshney, R.K, Graner, A., and Sorrells, M. E. (2005a). Genic microsatellite markers in plants: Features and applications. *Trends Biotechnol.* **23**, 48–55.

Varshney, R.K, Graner, A., and Sorrells, M. E. (2005b). Genomics-assisted breeding for crop improvement. *Trends Plant Sci.* **10**, 621–630.

Varshney, R. K., Prasad, M., Kota, R., Sigmund, R., Börner, A., Valkoun, J., Scholz, U., Stein, N., and Graner, A. (2005c). Functional molecular markers in barley: Development and applications. *Czech. J. Genet. Plant Breed.* **41**, 128–133.

Varshney, R. K., Sigmund, R., Korzun, V., Boerner, A., Stein, N., Sorrells, M., Langridge, P., and Graner, A. (2005d). Interspecific transferability and comparative mapping of barley EST-SSR markers in wheat, rye and rice. *Plant Sci.* **168**, 195–202.

Varshney, R. K., Balyan, H. S., and Langridge, P. (2006a). Wheat. *In* "The Genomes: A Series on Genome Mapping, Molecular Breeding and Genomics of Economic Species, Vol. 1, Cereals & Millets" (C. Kole, ed.), pp. 79–134. Springer, Dordrecht, The Netherlands.

Varshney, R. K., Grosse, I., Haehnel, U., Siefken, R., Prasad, M., Stein, N., Langridge, P., Altsch-mied, L., and Graner, A. (2006b). Genetic mapping and BAC assignment of EST-derived SSR markers shows non-uniform distribution of genes in the barley genome. *Theor. Appl. Genet.* **113**, 239–250.

Vassos, E. J., Barr, A. R., and Elington, J. K. (2003). Genetic conversion of feed barley varieties to malting types. *In* "Proceedings of Barley Technical/Cereal Chemistry Conference." Tasmania, Australiawww.cdesign.com.au/bts2005/pages/Papers_2003/papers/087VassosE.pdf.

Velculescu, V. E., Zhang, L., Vogelstein, B., and Kinzler, K. W. (1995). Serial analysis of gene expression. *Science* **270**, 484–487.

Wall, J. D., and Pritchard, J. K. (2003). Assessing the performance of the haplotype block model of linkage disequilibrium. *Am. J. Hum. Genet.* **73**, 502–516.

Wang, M. L., Atkinson, M. D., Chinoy, C. N., Devos, K. M., and Gale, M. D. (1992). Comparative RFLP-based genetic maps of barley chromosome-5 (1H) and rye chromosome-1 R. *Theor. Appl. Genet.* **84**, 339–344.

Weng, Y., and Lazar, M. D. (2002). Comparison of homoeologous group-6 short arm physical maps of wheat and barley reveals a similar distribution of recombinogenic and gene-rich regions. *Theor. Appl. Genet.* **104**, 1078–1085.

Yu, G. X., Bush, A. L., and Wise, R. P. (1996). Comparative mapping of homoeologous group 1 regions and genes for resistance to obligate biotrophs in *Avena*, *Hordeum*, and *Zea mays*. *Genome* **39**, 155–164.

Yu, J. K., Dake, T. M., Singh, S., Benscher, D., Li, W., Gill, B. S., and Sorrells, M. E. (2004a). Development and mapping of EST-derived simple sequence repeat markers for hexaploid wheat. *Genome* **47**, 805–818.

Yu, J. K., La Rota, M., Kantety, R. V., and Sorrells, M. E. (2004b). EST derived SSR markers for comparative mapping in wheat and rice. *Mol. Genet. Genomics* **271**, 742–751.

Zamir, D. (2001). Improving plant breeding with exotic genetic libraries. *Nat. Rev. Genet.* **2**, 983–989.

Zhang, D., Choi, D. W., Wanamaker, S., Fenton, R. D., Chin, A., Malatrasi, M., Turuspekov, Y., Walia, H., Akhunov, E. D., Kianain, P., Otto, C., Simons, K., *et al.* (2004a). Construction and evaluation of cDNA libraries for large-scale expressed sequence tag sequencing in wheat (*Triticum aestivum* L.). *Genetics* **168**, 595–608.

Zhang, H., Sreenivasulu, N., Weschke, W., Stein, N., Rudd, S., Radchuk, V., Potokina, E., Scholz, U., Schweizer, P., Zierold, U., Langridge, P., Varshney, R. K., *et al.* (2004b). Large-scale analysis of the barley transcriptome based on expressed sequence tags. *Plant J.* **40**, 276–290.

Zhou, W. C., Kolb, F. L., Bai, G. H., Domier, L. L., Boze, L. K., and Smith, N. J. (2003). Validation of a major QTL for scab resistance with SSR markers and use of marker-assisted selection in wheat. *Plant Breed.* **122**, 40–46.

Zhu, X., Zhang, S., Zhao, H., and Cooper, R. S. (2002). Association mapping using a mixture model for complex traits. *Genet. Epidemiol.* **23**, 181–196.

Index

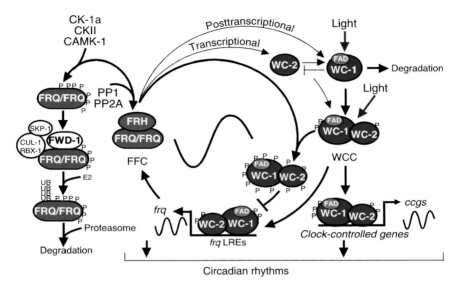

Chapter 2, Figure 2.1. Schematic drawing of the core circadian feedback loops and the associated degradation pathways as far as they are known. The central circadian oscillator is a transcription–translation-based feedback circuit in which a hypophosphorylated WC-1:WC-2 heterodimeric complex (WCC) binds to the frq promoter and activates frq transcription. Following its translation, FRQ dimerizes and forms a complex with an FRQ-directed RNA helicase (FRH) called the FFC. The FFC mediates WCC hyperphosphorylation, possibly by recruiting a kinase to the WCC, thus rendering the WCC transcriptionally inactive and closing the negative feedback loop. The fact that repressor activity of FFC requires proper FRQ phosphorylation by CKII has been omitted from this cartoon for reasons of simplicity. At the transcriptional and posttranscriptional level, FFC stimulates the production of WC-1 and WC-2. Similarly, WC-1 and WC-2 negatively and positively regulate each others levels. The mechanisms of these regulations are unknown but these additional regulatory loops are thought to make the central oscillator more robust. Light can enter the system via the WC-1 protein (or the WCC) that binds the chromophore FAD and acts as the main blue-light photoreceptor in Neurospora. The WCC also binds to downstream targets of the circadian oscillator and thus may confer rhythmicity and light responses of clock-controlled output genes, which ultimately help to control overt circadian rhythmicity. Timely degradation of the FFC is critical for maintaining robust rhythmicity and is controlled at the level of FRQ phosphorylation by the kinases CK-1a, CKII, and CAMK-1 and phosphatases PP1 and PP2A. Phosphorylation targets FRQ for degradation and is mediated via the ubiquitin/proteasome. FRQ binds transiently to FWD-1, an F-Box/WD40 repeat-containing adaptor that serves as the substrate recruiting subunit of the SCF-type ubiquitin (E3) ligases. Other components of the SCF complex are SKP-1, Cullin, and RBX-1, a RING domain-containing protein. On binding to the SCF complex, FRQ becomes quickly ubiquitinated and degraded by the proteasome. The stability of the SCF complex is regulated by the COP9 signalosome (not shown), adding another layer of regulation to FRQ degradation. For more detail see text.

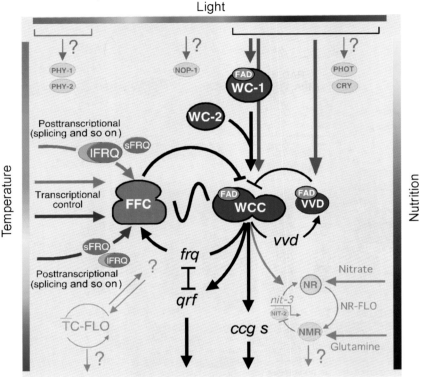

Circadian rhythmicity and light responses

Chapter 2, Figure 2.2. Schematic representation of the Neurospora circadian system with emphasis on the environmental input (light, temperature, nutrition) pathways that feed into the core oscillator. Known pathways are depicted in bold and color, whereas components with unknown or controversial functions are depicted in light gray. A simplified version of the central feedback loop (as depicted in more detail in Fig. 2.1) is shown at the center of the system. Light perception in Neurospora is probably mediated by a number of photoreceptors that absorb wavelength across the entire spectrum of visible light, but only the WC-1 and VVD proteins are established as blue-light photoreceptors, which seems to mediate all known light responses in Neurospora. WC-1 is the primary photoreceptor and is required for the expression of VVD, which the serves as a secondary photoreceptor and repressor of WCC activity. The VVD feedback loop mutes light responses in Neurospora and impacts on light resetting and entrainment of the Neurospora circadian clock (for more information see text). Similarly, the expression of *qrf*, an antisense transcript to the *frq* mRNA, mutes light resetting of the Neurospora clock via an unknown mechanism. Temperature is sensed via transcriptional and posttranscriptional responses that act on the *frq* gene and its products. At higher temperatures, more large FRQ (lFRQ) is made, whereas at lower temperatures, relatively more small FRQ (sFRQ) is present. The overall quantity of FRQ and the ratios of lFRQ to sFRQ are largely regulated at the posttranscriptional level via temperature-sensitive alternative splicing mechanisms and temperature-dependent ribosome scanning mechanisms (see text for more detail). The environmental signals are integrated at the level of the oscillator and control a large number of

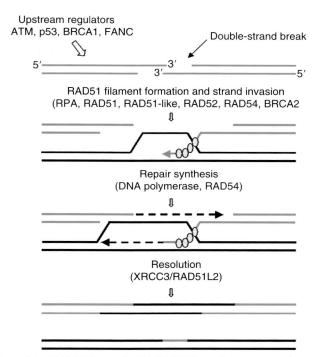

Upstream regulators
ATM, p53, BRCA1, FANC

Double-strand break

5'——————————————3'   ——————————————
————————————————3'———————————————————5'

RAD51 filament formation and strand invasion
(RPA, RAD51, RAD51-like, RAD52, RAD54, BRCA2

⇓

Repair synthesis
(DNA polymerase, RAD54)

⇓

Resolution
(XRCC3/RAD51L2)

⇓

Chapter 3, Figure 3.1. DNA DSB repair by HR. RAD51 forms filaments on single-stranded 3' DNA after displacement of single-stranded DNA-binding protein RPA and catalyzes strand transfer between broken sequence and its homologue (strand invasion) to allow resynthesis of the damaged DNA fragment (Krogh and Symington, 2004). RAD52 and probably RAD51-like proteins assist displacement of RPA and annealing of the invaded strand to the undamaged complementary homologous strand. After repair synthesis, Holliday junctions are resolved possibly by XRCC3/RAD51L2 heterodimer. RAD54 and RAD51-like proteins seem to play a role before and after strand invasion to facilitate opening up of DNA and removal of proteins bound to single-stranded DNA. The role of HR regulators is less understood. Some of these proteins are involved in upstream steps of DNA damage response, while BLM and WRN appear to be involved in branch migration of Holliday junction, as discussed below.

clock-controlled genes (*ccgs*) that impart rhythmicity on the physiology of the whole organism. Little is known about nutritional input into the circadian oscillator. A few FRQ-less oscillators that become visible or operate in the absence of the circadian FRQ-WCC oscillator have been described of which only two are depicted here. The nitrate-reductase FLO (NR-FLO) is predicted to center around the pathways of nitrate assimilation and only present when nitrate is the sole nitrogen source but is suppressed by ammonium or glutamine. The components of a temperature-controlled FLO (TC-FLO) that regulates rhythmic conidiation in temperature cycles are unknown. Whether these FLOs are controlled by circadian oscillators and contribute to overt circadian rhythmicity in Neurospora has not been established or is controversial.

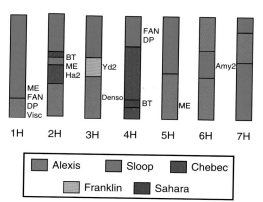

Chapter 5, Figure 5.1. Genetic ideotype based around a set of five barley varieties. Each chromosome has been colored to reflect the desired region. The location of some key loci is indicated (ME, malt extract; FAN, free amino nitrogen; DP, diastatic power; Visc, wort viscosity; BT, boron tolerance; Ha2, cereal cyst nematode resistance; Yd2, barley yellow dwarf virus resistance; Denso, dwarf; Amy2, α-amylase) [adapted from Jefferies (2000)].